Applied Descriptive Geometry

Applied Descriptive Geometry

Susan A. Stewart

DELMAR PUBLISHERS INC.

Delmar Staff
Administrative editor: Mark W. Huth
Production editor: Barbara A. Christie

For information address Delmar Publishers Inc.,
2 Computer Drive West, Box 15-015,
Albany, New York 12212

Printed in the United States of America
Published simultaneously in Canada
by Nelson Canada,
a division of International Thomson Limited

10 9 8 7 6 5 4 3 2 1

Library of Congress Cataloging in Publication Data

Stewart, Susan A. (Susan Ann), 1947-
    Applied descriptive geometry.

    Includes index.
    1. Geometry, Descriptive. I. Title.
QA501.S77 1986   516   85-10395
ISBN 0-8273-2377-8

CONTENTS

CHAPTER
SIX

INTERSECTING AND NONINTERSECTING LINES

CHAPTER
SEVEN

PLANE RELATIONSHIPS

CHAPTER

ELEVEN

VECTOR GEOMETRY

339

CHAPTER
TWELVE

MINING AND CIVIL ENGINEERING APPLICATIONS 375

PREFACE

PURPOSE

Descriptive geometry and orthographic projection are the graphic tools of the engineering function. This book has been developed to teach college engineering and technology students the fundamental concepts of descriptive geometry through an emphasis on logical reasoning, visualization, and practical applications.

FORMAT

The book is constructed in a text/workbook format. Each chapter includes relevant instructional concepts and examples, a chapter test, and a selection of applied problems. A portion of the problems may be solved directly on the worksheets, while others require solution on vellum, or other drawing media. The chapter tests may be answered directly on the workbook pages, removed from the workbook, and submitted for evaluation.

HIGHLIGHTS

For easy reference and emphasis, important concepts appear in boldface type. Also in bold type are the notations alphabetically identifying points, lines and planes.

LENGTH OF COURSE

The course content is designed to be used in an eleven to sixteen week time frame, with variability resulting from the quantity of problem assignments, the depth of classroom discussion, and instructor's prerogative. Generally speaking, each chapter can be covered in a one-hour lecture.

COURSE PLAN

Most courses in descriptive geometry are based on the principles of orthographic projection and four fundamental views: (1) true length of a line, (2) point view of a line, (3) edge view of a plane, and (4) true shape of a plane. This text begins with these four fundamentals, proceeds to line relationships, to plane relationships, and to various other important concepts. It progresses in small steps from the simple to the more complex. However, the instructor had some flexibility in choosing the order of the topics. The block diagram on the following page indicates the interdependence of the chapters.

Orthographic projection and the visual orientation for descriptive geometry are presented in Chapters 1 and 2. The concepts of auxiliary views, both primary and secondary, are discussed in Chapter 3, in conjunction with the first two fundamental views. The remaining two fundamental views are treated in Chapter 4, completing the core concepts section of the text.

Line relationships are explained in Chapters 5 and 6, providing the foundation for a discussion of plane relationships in Chapters 7 and 9. The concept of revolution as an alternative method of problem solution is presented in Chapter 8, and applied specifically to sheet metal developments in Chapter 10. Chapter 11, Vector Geometry, and

Chapter 12, Mining and Civil Engineering Applications, provide the student with two additional and very different applications for the tools of descriptive geometry.

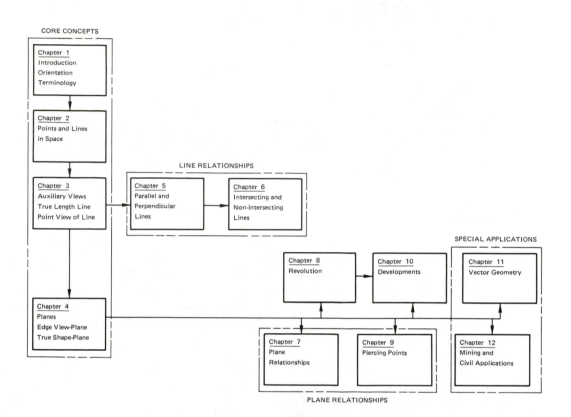

APPLICATIONS

Special emphasis has been made throughout the text on applications in the various engineering disciplines. Examples have been taken from mechanical, industrial, piping, aerospace, marine, civil, structural, and architectural applications. The value of descriptive geometry can be readily seen by students as they apply the tools and techniques to practical problems.

INSTRUCTOR'S MANUAL

The Instructor's Manual contains answers to all chapter problems and chapter tests. In addition, the manual includes instructional objectives for each chapter.

ACKNOWLEDGMENTS

Since its inception, this project has undergone extensive modifications and improvements. The following professors reviewed the manuscript and provided valuable suggestions for its improvement: John H. Crawford, John Denison, Delmas Green, and Jack Koller. I am greatly indebted to my illustrator, Stacy Thayer, for her talent, for many useful discussions of both the content and the illustrations, and for her constant encouragement. I wish to thank also my colleagues at Clackamas Community College, Michael Durrer, Terence Shumaker, and David Madsen for their encouragement and support, and for many helpful suggestions. My thanks are also due to Joni Bishop, who helped me prepare the final version of the Instructor's Manual.

I am deeply grateful to the many students who have taken my descriptive geometry courses. Their enthusiasm and interest were the driving force behind the development of this book. Their comments on earlier drafts provided the most valuable reviews of all.

Susan A. Stewart

CHAPTER
ONE

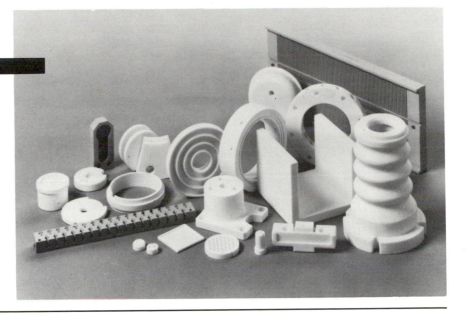

Orthographic Projection and Descriptive Geometry

1.1 INTRODUCTION

For many years, we have recognized that we live in a world of three-dimensional objects. The problem encountered in drawing three-dimensional objects is one of accurately describing the object on a two-dimensional sheet of paper. A photograph of a pictorial drawing shows how an object appears well enough, but generally it is not adequate for transmitting the details of the idea from the designer to the manufacturer. Before an object can be made, an exact description must be conveyed to the maker. This is the purpose of orthographic projection and descriptive geometry.

Today orthographic projection and descriptive geometry are among the most valuable subjects in technical and engineering education. The fundamentals of descriptive geometry are based on the principles of orthographic projection. These types of drawings are practically indispensible to the engineer. They are the language of the engineering profession. Some drawings use orthographic projection to transmit an exact description of an object to the maker, while other drawings use descriptive geometry and orthographic projection to make a type of layout drawing that aids in the solution of a spatial problem. So orthographic projection and descriptive geometry are not different subjects really, but rather they are different areas within the same subject that make use of the same tools and thought processes.

Descriptive geometry differs from orthographic projection in that it is not necessarily concerned with a finished drawing that communicates between the designer and the fabricator, but instead is concerned with a study of spatial relationships that occur during the design process. A spatial relationship could be the relationship of one part of a structure to another, of one part of a mechanism to another part of the mechanism, of one surface of a part to another surface on the same part, or of a highway surface to its adjacent surroundings. The examples are endless. Using the tools of descriptive geometry, the study of these relationships is a process of drawing the objects or situations in a series of successive auxiliary views until the required solution is found.

1.2 ORTHOGRAPHIC PROJECTION FUNDAMENTALS

1.2.1 Orientation

In an orthographic drawing, it is common practice to draw different views of an object. To do this the drafter must first imagine the object is placed in a definite position which is usually a normal or natural position for the object. Once the object has been placed, it remains in that position. If you wish to see a different view, you simply imagine that you have moved around it in order to view the object from another position in space. In this method, the observer changes position, not the object. In order to more fully understand and to use this system of projection, several definitions must be learned.

1.2.2 Definitions

The following elements are illustrated in Figure 1-1, which is a pictorial drawing of all seven definitions. Study the illustration and definitions carefully.

Figure 1-1. Pictorial drawing of descriptive geometry terminology.

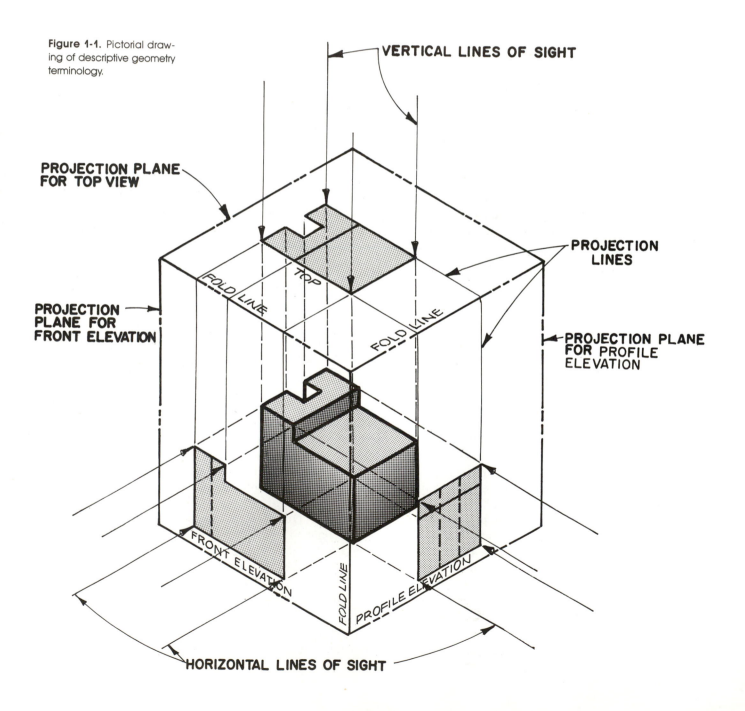

1. ORTHOGRAPHIC PROJECTION (DRAWING) means **right-angle** projection. It is a method of drawing which uses parallel lines of sight at right angles (90°) to a projection plane.
2. A LINE OF SIGHT is an imaginary straight line from the eye of the observer to a point on the object. All lines of sight for a particular view are assumed to be parallel, which means that the observer is either an infinite distance away, or that the observer occupies a slightly different position when looking at each point on the object.
3. The PROJECTION PLANE is an imaginary surface on which the view of the object is projected and drawn. This surface is imagined to exist between the object and the observer. Your (the observer's) lines of sight are **always** perpendicular to the projection plane.
4. PROJECTION LINES are straight lines at 90° to the fold lines or reference lines, which connect the projection of a point in a view to the projection of the same point in an adjacent view. These lines are required for obtaining views, but they are rarely shown on a finished drawing.
5. A FOLD LINE, or REFERENCE LINE, is the line of intersection between two projection planes. It is the line on which one plane is folded to bring it into the plane of the adjacent projection plane. A fold line appears as an edge view of a projection plane when that projection plane is folded back 90° from the plane being viewed.
6. A TOP, or HORIZONTAL VIEW, is an orthographic view for which the lines of sight are vertical and for which the projection plane is level.
7. An ELEVATION VIEW is an orthographic view for which the lines of sight are horizontal. The projection plane is vertical in an elevation view.

1.3 MULTIVIEW PROJECTION

Now that you have studied Figure 1-1, let's move on to the type of orthographic projection called multiview projection, or multiview drawing. **The primary purpose of a multiview drawing is to obtain views of an object on which actual measurements can be shown.** Refer again to Figure 1-1, and notice that the front surface is positioned parallel to the projection plane showing the front elevation. The measurements on that plane are true.

In the standard orthographic projection system the projection planes used form an imaginary glass box which surrounds the object. **Each of the six planes of this box is perpendicular to any plane next to it.** The three mutually perpendicular projection planes used most often are called the primary, or principal, planes of projection. The primary planes are the horizontal (top) projection plane, the front projection plane, and the profile (side) projection plane. **For each plane the point of sight is infinity, and the projection lines are perpendicular to the projection plane.**

Figure 1-2 illustrates orthographic projection. Notice again that the object is located so that its major axes are parallel or perpendicular to the three primary planes. Additional planes, called auxiliary planes, are often required to show all of the necessary details of the object. Auxiliary planes will be discussed in Chapter 3.

Figure 1-2 also illustrates the labeling of the primary fold lines. The fold line between the horizontal and front projection planes is labeled **H/F**. When you look at the front plane, the fold line labeled H/F represents an edge view of the horizontal plane. When you change position and look at the horizontal plane, the fold line labeled H/F represents an edge view of the front projection plane. The fold line between the front and the profile projection plane is labeled **F/P**. When viewing the front plane, the profile plane appears as an edge, represented by fold line F/P. How does the front-projection plane appear when you are looking at the profile projection plane?

1.4 UNFOLDING THE PROJECTION PLANES

All multiview drawings represent a standard arrangement of orthographic views drawn on a single plane (the sheet of paper). You have learned about these imaginary planes, inside of which sits the object, and about the three views of the object projected onto the

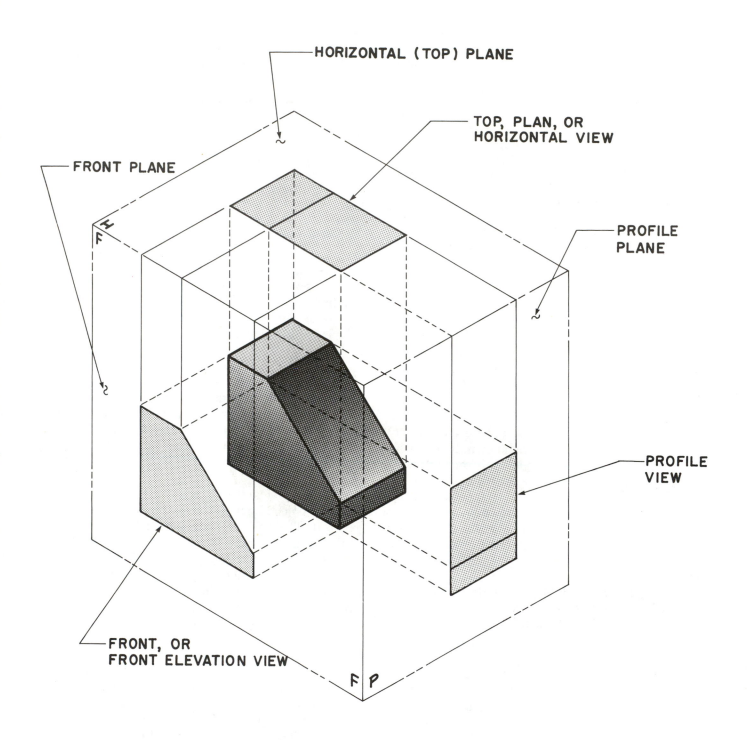

Figure 1-2. Pictorial of orthographic projection.

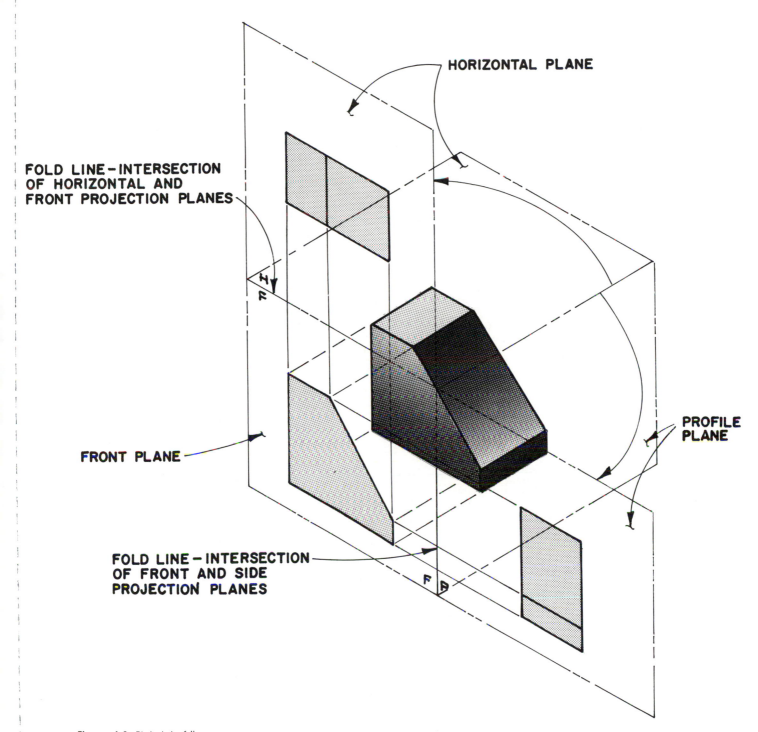

HORIZONTAL PLANE

FOLD LINE – INTERSECTION
OF HORIZONTAL AND
FRONT PROJECTION PLANES

H
F

FRONT PLANE

PROFILE
PLANE

FOLD LINE – INTERSECTION
OF FRONT AND SIDE
PROJECTION PLANES

F P

Figure 1-3. Pictorial of the
orthographic box unfolded.

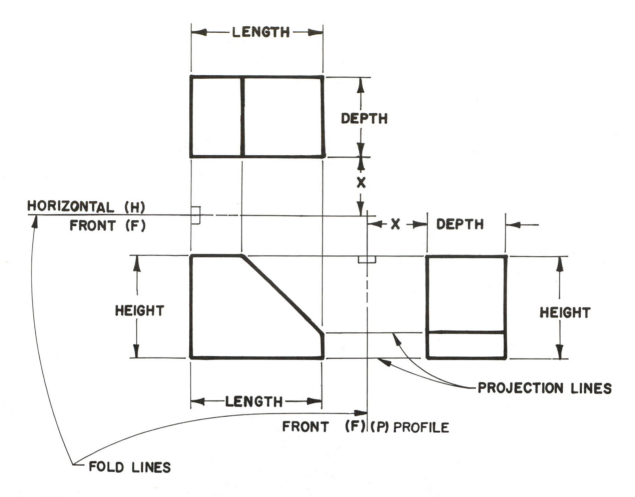

Figure 1-4. Multiview drawing.

planes. To make this system useful, each projection drawn must be shown in one plane (the plane of the paper).

First, imagine that the front-projection plane remains stationary and the other two planes are unfolded (rotated) about fold lines H/F and F/P respectively. Figure 1-3 is a pictorial illustrating this way of unfolding the planes. Notice again that the fold lines representing the intersection of the front and the horizontal planes is labeled H/F, and that the fold line between the front and profile planes is labeled F/P.

The resulting multiview drawing of the object is shown in Figure 1-4. Study this illustration carefully. Notice which dimensions are the same, and that the top and side views are equidistant from their respective fold lines. Why do you think this occurs?

Special attention should be paid to the following facts:

- The **length** of the object is evident in both the top and the front views.
- The **depth** of the object is seen in both the top and the profile (side) views.
- The **height** of the object is seen in the front and the profile (side) elevations.
- The projection lines are always perpendicular to the fold lines that they cross.
- The two distances **X** are equal. **X** equals the distance from the front-projection plane to the front surface of the object.

1.5 FOLD LINES

Most *finished* engineering drawings do not include fold lines. In fairly simple drawings which require only two or three views of any object, it is good industrial practice to omit the fold lines. However, in more complicated drawings and problems approached with the tools of descriptive geometry, it is important that there be some place from which to measure and a standard way of measuring. The purpose of fold lines is to provide this means of measuring and reference. Therefore, they will be used throughout this text. It is expected that the student will use them on all the chapter problems.

1.6 ADJACENT VIEWS

Views are adjacent, or related to each other when:

- Their projection planes are perpendicular to each other in space.
- They have fold lines between them.
- The two views of any point lie on the same projection line, which is at 90° to the fold line between the views.

Figure 1-5 shows a pictorial and orthographic drawing demonstrating these three conditions of adjacent views.

Figure 1-5. Adjacent views.

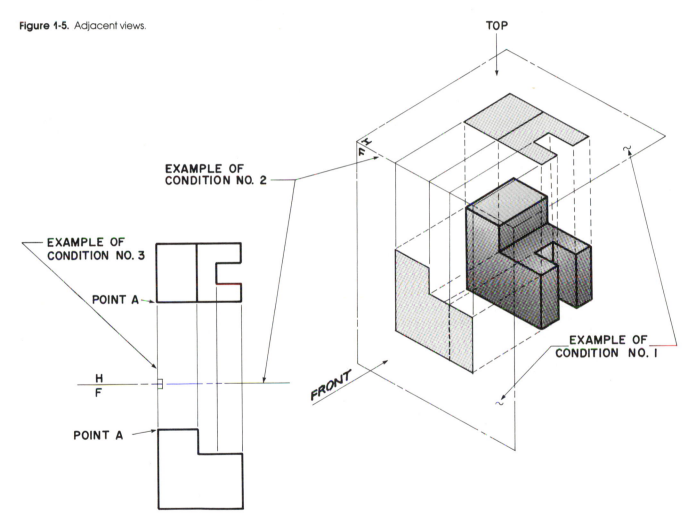

CHAPTER PROBLEMS

The problems contained in this text are meant to provide practice in the use of descriptive geometry principles. They are designed to test your understanding of the basic fundamentals and of the theoretical methods used, and to teach you to apply the principles while finding solutions to a large variety of engineering problems. Many of the problems are derived from the broad spectrum of engineering settings. Each problem should be solved as accurately and neatly as possible.

Problem 1: Given a pictorial drawing of a **Slide Bracket,** draw the top, front, and right side views in their conventional relation- ship, as indicated by the fold lines. Scale the pictorial drawing to obtain the proper dimensions.

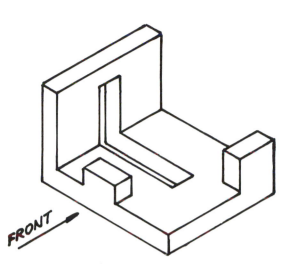

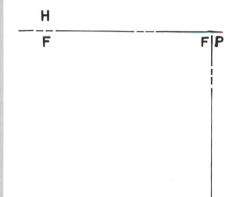

H

F   F|P

Problem 2: Given a pictorial drawing of a **Locator Base,** draw the top, front, and right side views in their conventional relation- ship, as indicated by the fold lines. Scale the pictorial drawing to obtain the correct dimensions.

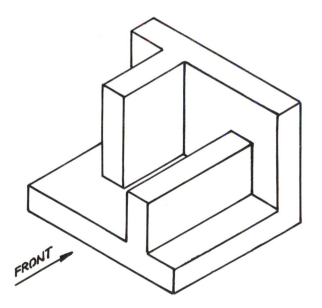

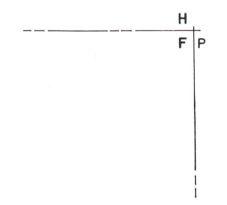

H

F|P

DR. BY:	COURSE & SEC:	SCALE:	DATE:

Problem 3: Given the pictorial drawing of a **Stop Block,** draw the top, front, left side, and right side views in their conventional relationship. Scale the pictorial drawing for the correct dimensions. Use fold lines and label them appropriately.

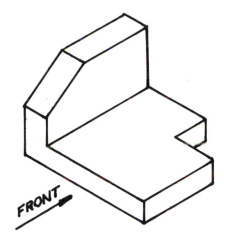

Problem 4: Given the pictorial drawing of a **Wedge Block,** draw the top, front, and right side views. Scale the pictorial drawing for the correct dimensions. Use fold lines and label them appropriately.

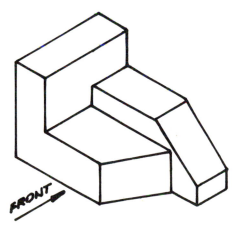

DR. BY:	COURSE & SEC:	SCALE:	DATE:

Problem 5: Given the pictorial drawing of a **Clip,** draw the top, front, and right side views, including the necessary fold lines, label correctly. Scale the pictorial for the correct dimensions.

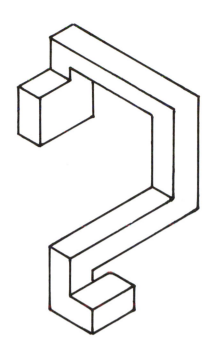

Problem 6: Given the top and right side views of a **Guide Block,** draw the front view.

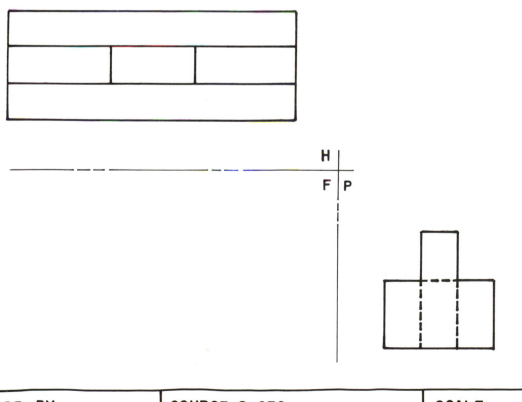

H

F | P

| DR. BY: | COURSE & SEC: | SCALE: | DATE: |

Problem 7: Given the front and the right side views of the **Index Pin,** draw the top view. Include the H/F fold line.

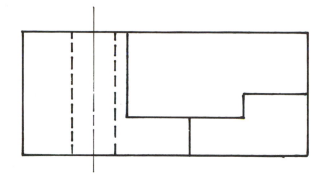

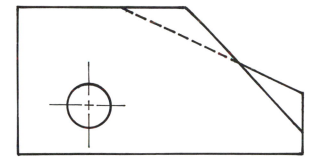

Problem 8: Given the top and the front views of the **Shift Fork,** draw the right side view. Include the F/P fold line.

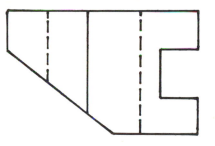

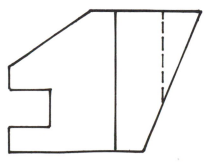

| DR. BY: | COURSE & SEC: | SCALE : | DATE: |

Problem 9: Given the profile and the front views of the **Locator Bracket,** draw the top view. Include the fold lines between the adjacent views, correctly labeled.

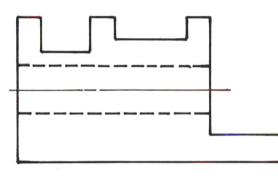

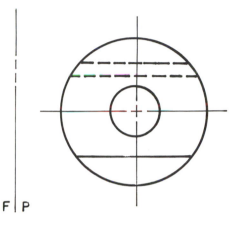

F | P

Problem 10: Given the top and the front views of the **Stop Block,** add the right side view. Include the fold line between the adjacent views, correctly labeled.

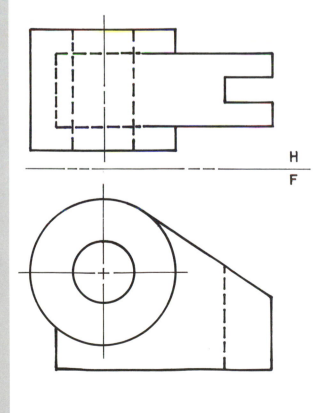

H
F

| DR. BY: | COURSE & SEC: | SCALE : | DATE: |

CHAPTER ONE

TEST

1. Define the following:

   a. Orthographic projection:

   b. Projection planes:

   c. Fold line:

   d. Line of sight:

   e. Projection lines:

2. What is the primary purpose of a multiview drawing?

3. Why are the projection planes unfolded?

4. In a standard multiview drawing, including the three primary views, (a) which two views will show the height of the object viewed?, and (b) which two views will show the overall length of the object?

   (a) _____

   (b) _____

5. What relationship do the projection lines have to the fold lines?

6. What is the purpose of a fold line?

7. What three conditions are necessary for views to be adjacent to one another?

8. Examine Figure 1-4 in this chapter and explain why the two distances denoted X are equal.

Points and Lines in Space

1.1 INTRODUCTION

You are now well versed in the methods of representing three-dimensional solids on a single sheet of paper. In actuality, these solids are developed from a series of connected points. Two points can define a line. Lines can define surfaces. Finally, surfaces combine to form an object. Since it is so fundamental to the application of descriptive geometry to technical problems, it is important to study in detail how points and lines may be shown in orthographic projection.

2.2 POINTS IN SPACE

In theory a point has location only and no dimensions. For location and accuracy the points in this text will be drawn using a small cross. Figure 2-1 is a pictorial drawing of a wedge block in the orthographic *box*, including three views projected onto the principal planes. Point **A** is identified on the wedge block in space, as well as in all three projection planes. Since the top and the right-profile projection planes are both perpendicular to the front plane before they are rotated into the front plane, the top and the profile projections of point **A** are the same distance, **X**, from the front plane. This distance is measured from the fold lines H/F and F/P. Notice that the point in space is indicated by the capital letter **A**, while each view is identified by the lowercase letter plus the relevant subscript a_H, a_F, a_P.

Figure 2-2 is a multiview drawing of the wedge block that results when the views are rotated into the plane of the front-projection plane as discussed in Chapter 1. Again the distance **X** is indicated as the distance from the front plane to point **A**. It represents the distance that point **A** lies behind the front plane, in both the top and the right-profile views.

To simplify matters, Figure 2-3 illustrates only point **A** in the three views, which also denote distance **X**. Study this figure carefully. Distance **Y** is also dimensioned and represents the distance that point **A** lies below the horizontal projection plane. Try to visualize this situation.

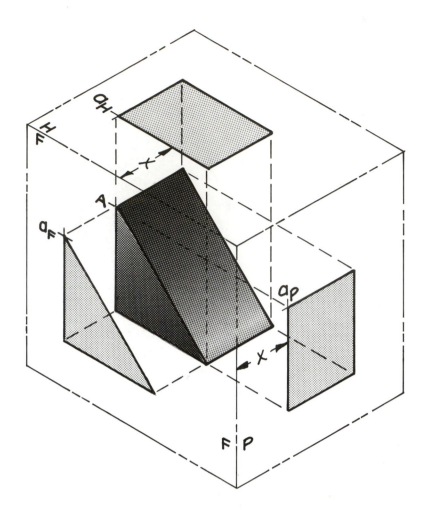

Figure 2-1. Pictorial drawing of a wedge block in the orthographic box.

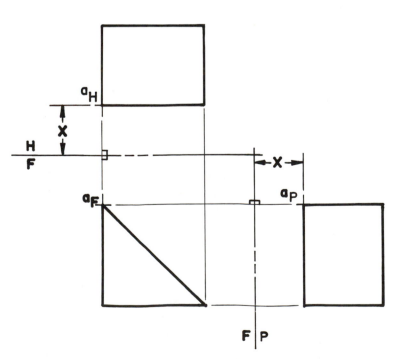

Figure 2-2. Multiview drawing of a wedge block.

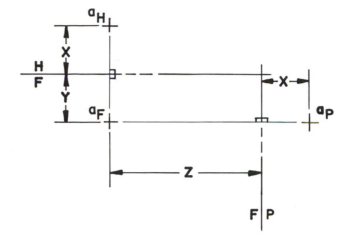

Figure 2-3. Three views of
point A.

2.2.1 Locating a Point

In Figure 2-4 you are given the fold lines representing the intersection between the horizontal and front projection planes (fold line H/F), and the front and the right-profile projection planes (fold line F/P). You are also given the top and the front views of point **M** on the step block. Using the concepts discussed thus far, accurately locate point **M** in the right side view. In addition, determine the distance in inches for the following:

- Point **M** below the horizontal-projection plane.
- Point **M** behind the front-projection plane.
- Point **M** to the left of the profile-projection plane.

The procedure for solving the problem in Figure 2-4 is as follows:

1. Draw a projection line from m_F across fold line F/P and perpendicular to F/P.
2. Measure the distance from fold line H/F to m_H, which represents the distance that point **M** lies behind the front projection plane.
3. Transfer the distance measured to the profile view from the fold line F/P on the projection line drawn from the m_F. Please keep these concepts in mind, as they are used in orthographic drawing constantly. The fold lines are used as reference lines (lines from which to measure) for the location of points.

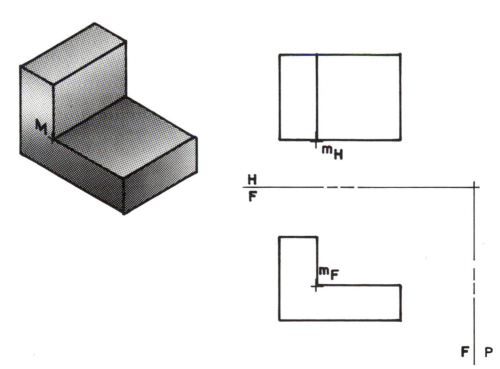

Figure 2-4. Exercise: Locating point M in the profile view.

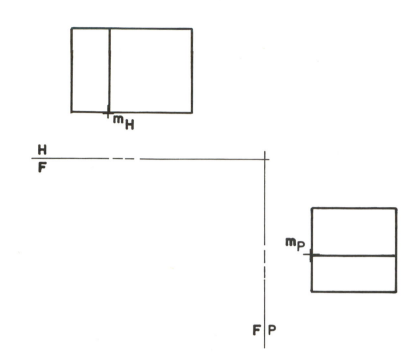

Figure 2-5. Exercise: Locating point M in the front view.

Let's try viewing the same object. But this time in Figure 2-5 you are given the top and the right side views of point **M** on the step block. In this example, accurately locate Point **M** in the front view.

The procedure for solving this problem is as follows:

1. Draw a projection line from m$_H$ that is perpendicular to fold line H/F and extends into the front view. This represents the distance **M** is located to the left of the right-profile plane. You are transferring this distance from the top to the front view.
2. Draw a projection line from m$_P$ that is perpendicular to fold line F/P and extends into the front view. Here you are transferring the distance that point **M** lies below the top plane into the front view.
3. When the two projection lines intersect, you have located the front view of point **M** (m$_F$).

2.3 LINES IN SPACE

Since a line can be two points which are connected, the two end points of a line may be projected onto each projection plane to form the views of that line. Figure 2-6 is a pictorial drawing of the orthographic projection of a wedge, including the top, front, and profile projections. Line **CD** is indicated on the object and in all three views. The multiview drawing of the wedge is shown in Figure 2-7. Study both of these figures and be aware that the length of line **CD** is not the actual, or true, length (T.L.) in any of these views. In each view the line appears slightly shorter than it really is and is called **foreshortened**.

With the help of Figure 2-8, try to visualize the position of line **CD** in space. Notice that distances **X** and **Y** are the same in the top and the profile views.

Examine the wedge shown in Figure 2-9. Pay special attention to lines **AB, AC, CD,** and **AD** and the views of these lines projected onto the three projection planes. Line **AC** is vertical, that is, straight up and down in relation to the top projection plane. Because it is vertical it will appear as a point in the top view. Since it is actually parallel to the front plane, you will see its true length in the front view. In contrast, line **AB** is horizontal (level), that is, parallel to the horizontal projection plane. Because it is parallel to the horizontal plane, it appears true length in the top view. Notice that line **AB** appears as a point in the front plane. Line **CD** is parallel to both the horizontal and the front planes, therefore its true length is seen in both the top and the front views. Finally, line **AD** is parallel to the front plane, therefore is appears true length in the front view. Because it is

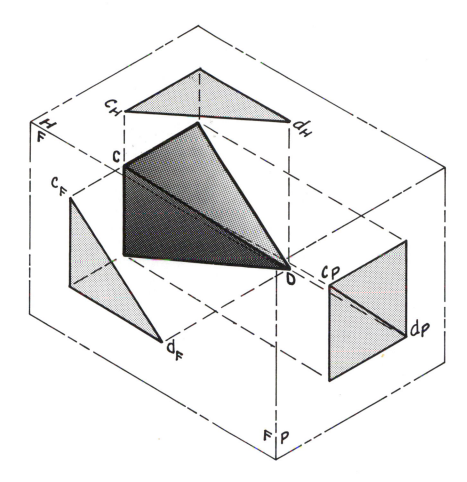

Figure 2-6. Pictorial of a wedge in the orthographic box.

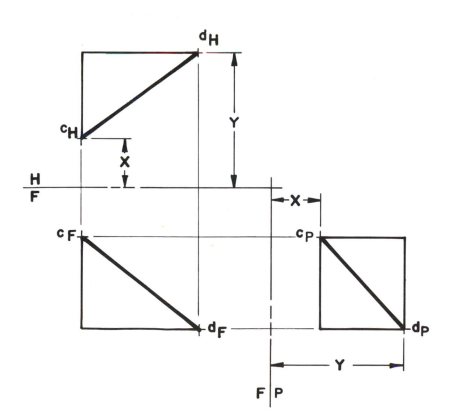

Figure 2-7. Multiview drawing of a wedge.

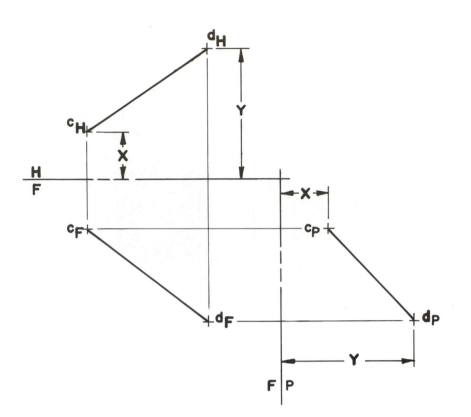

Figure 2-8. Three views of line CD.

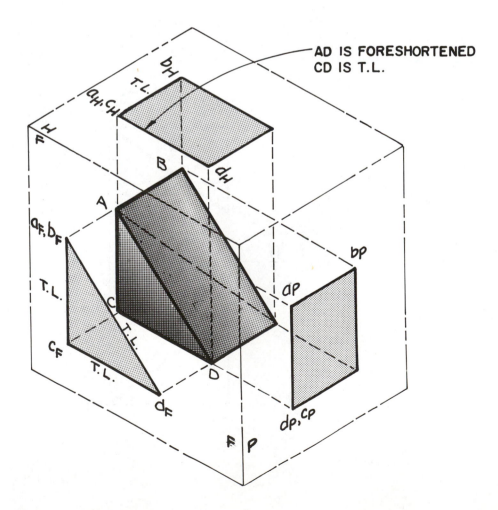

AD IS FORESHORTENED
CD IS T.L.

Figure 2-9. Pictorial of a wedge: True length lines.

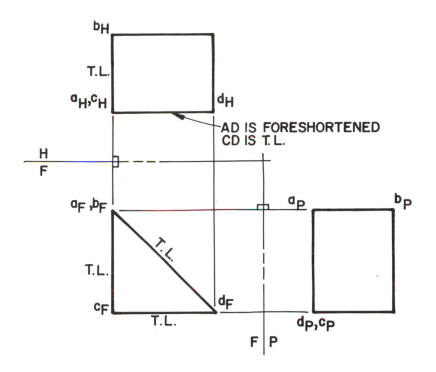

Figure 2-10. Multiview drawing: True length lines.

sloping, and not parallel to the top plane, it is foreshortened in the top view. How do lines **AB**, **AC**, **CD**, and **AD** appear in the profile view?

It is always true that a straight line will appear true length when it is drawn on a projection plane that is parallel to the line. The multiview drawing of the wedge block is shown in Figure 2-10. Take some time to visualize these lines and the various views. Ask yourself how each line appears in each view. If you practice such mental visualization regularly, even the most complex problems will seem much simpler.

2.4 TYPES OF LINES

A line is generally considered to be a straight line unless otherwise designated. In descriptive geometry, there are several line types, or classifications. Lines that are parallel to one of the principal projection planes (front, horizontal, or profile) are called principal lines.

A line that is parallel to the front projection plane is called a **frontal line**; and its projection will be true length in the front view. Figure 2-11 shows a pictorial and an orthographic drawing of a wedge. Line **AB** is a frontal line because it is parallel to the front plane and, therefore, is drawn true length in the front view.

When a line is parallel to the horizontal projection plane, it is called a **horizontal line**. Its projection will appear in true length in the top (horizontal) view. Figure 2-12 illustrates a wedge both pictorially and orthographically. Line **CD** is a horizontal, or level, line because it is parallel to the horizontal projection plane and, therefore, appears true length in the top view.

Finally, in the same manner, a **profile line** is one that is parallel to the profile projection plane. Figure 2-13 shows a pictorial and an orthographic drawing of another wedge. Line **EF** is a profile line because it is parallel to the profile projection plane. Both the top and the front views of a profile line are perpendicular to the H/F fold line. The true length of line **EF** appears in the profile view.

Keeping in mind that a line is simply two points connected, review the method for finding the location of points that you learned in Section 2.2.1. This time locate points **X** and **Y** in the profile view to form line **XY** in Figure 2-14.

The procedure for finding the profile view of line **XY** is as follows:

1. Draw projection lines from x_F and y_F, respectively, that are perpendicular to the fold line F/P. Now ask yourself, how far behind the front projection plane are points

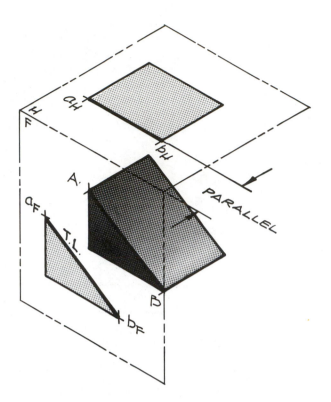

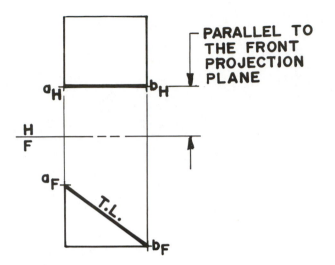

Figure 2-11. The frontal line.

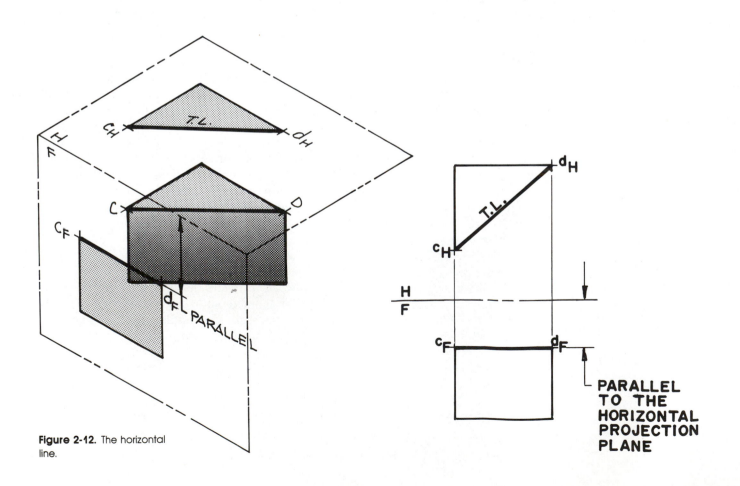

Figure 2-12. The horizontal line.

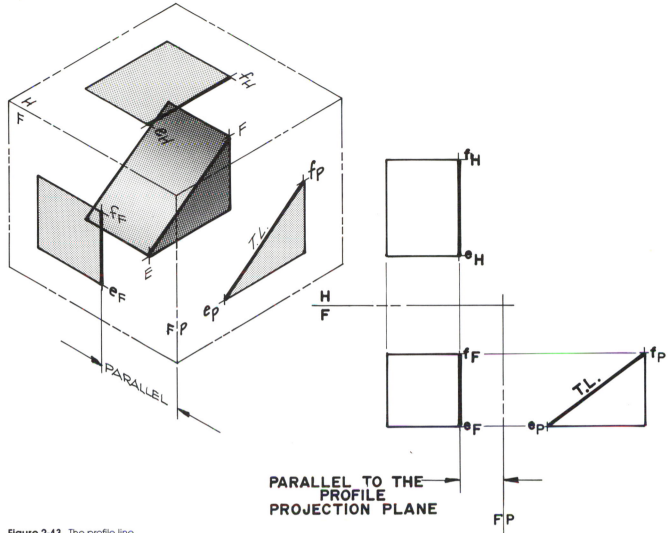

Figure 2-13. The profile line.

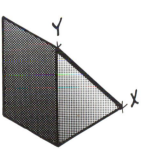

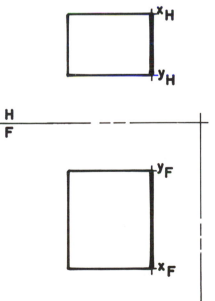

Figure 2-14. Exercise: Line location.

X and **Y**? Where do I find that information? If you knew the answer is the top view, you are doing fine.

2. Measure the distance from the fold line H/F to x_H and y_H. Transfer each of these measurements to the profile view by laying them out on the projection lines from x_F and y_F. This will locate x_P and y_P. Can you complete the profile view of this object?

Figure 2-15 includes a pictorial and an orthographic drawing of a locator block that illustrates the concepts discussed in this chapter. Lines **AB**, **BC**, and **AC** are singled out. Line **AB** is a horizontal line, because it is parallel to the horizontal projection plane. Since line **AC** is parallel to the front projection plane, it is a frontal line. Finally, line **BC** is a profile line, because it is parallel to the profile projection plane. Ask yourself the following questions:

1. How far behind the front projection plane are points **A**, **B**, and **C**, and where are these distances found?
2. How far below the horizontal projection plane are points **A**, **B**, and **C** and where are these distances found?

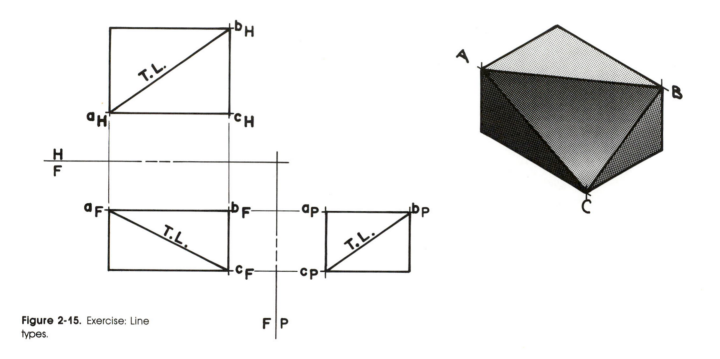

Figure 2-15. Exercise: Line types.

CHAPTER PROBLEMS

The problems in this chapter are more theoretical than most of the others in this text. Your ability to visualize points and lines as the building blocks of objects and other engineering situations is very important.

Problem 1: Points **A**, **B**, **C**, **D**, **E**, and **F** are located in space as shown by the given horizontal and frontal projections. Examine the points carefully, and answer the following questions.

a. Is point **A** above point **C**? _____

b. Is point **A** behind point **B**? _____

c. Is point **C** above point **D**? _____

d. Which point is closest to the front plane? _____

e. Which two points are the same elevation (height)? _____

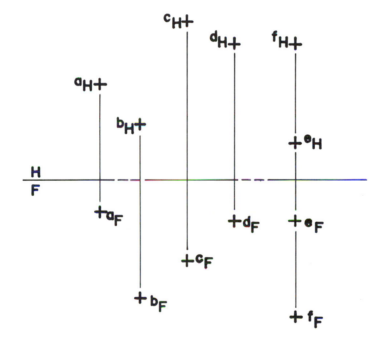

| DR. BY: | COURSE & SEC: | SCALE: | DATE: |

Problem 2: Given the horizontal and front views of the letter **A**, find the right profile view of the letter **A**. (Hint: do so by relocating points **A, B, C, D,** and **E**.)

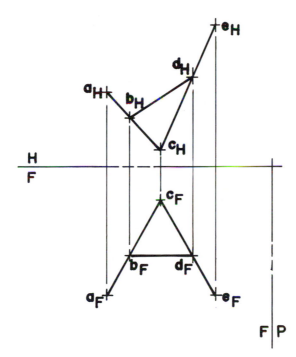

Problem 3: The horizontal and front projections of six lines are given below. Examine the lines and complete the statements found on the following page. (Note: When two types of line types apply, give both.)

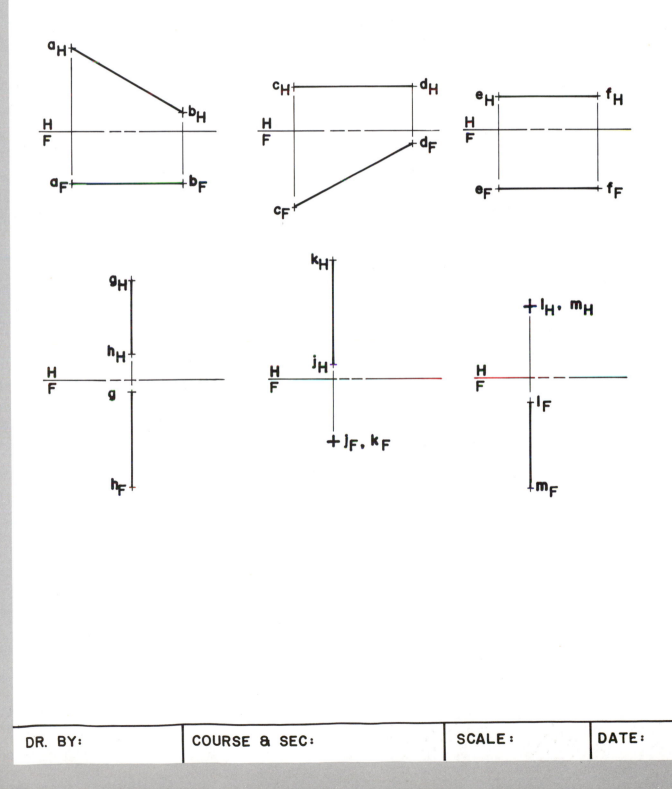

DR. BY:

COURSE & SEC:

SCALE:

DATE:

Problem 3: Complete the following statements based on your examination of the six lines given on the previous page. (Note: When two types of lines apply, give both.)

a. Line **AB** is parallel to the _____ plane.

b. Parallelism between line **AB** and the _____ plane is evident from the position of the _____ projection of line **AB**.

c. Line **AB** is in true length in the _____ projection.

d. Line **AB** is what type of line? _____

e. Line **CD** is parallel to the _____ plane.

f. Parallelism between line **CD** and the _____ plane is evident from the position of the _____ projection of line **CD**.

g. Line **CD** is true length in the _____ projection.

h. Line **CD** is what type of line? _____

i. Line **EF** is parallel to the _____ plane.

j. Parallelism between line **EF** and the _____ plane is evident from the position of the _____ projection of line **EF**.

k. Line **EF** is true length in the _____ projection.

l. Line **EF** is what type of line? _____

m. Line **GH** is parallel to the _____ plane.

n. Line **GH** is true length in the _____ projection.

o. Line **GH** is what type of line? _____

p. Line **JK** is parallel to the _____ plane.

q. Line **JK** is in true length in the _____ projection.

r. Line **JK** is what type of line? _____

s. Line **LM** is parallel to the _____ plane.

t. Parallelism between line **LM** and the _____ plane is evident from the position of the _____ projection of line **LM**.

u. Line **LM** is in true length in the _____ projection.

v. Line **LM** is what type of line? _____

Problem 4: In each the four problems below, there is a view which is missing. Complete the missing view. Label the true length view, and indicate the type of line for each problem.

TYPE OF LINE _____

TYPE OF LINE _____

TYPE OF LINE _____

TYPE OF LINE _____

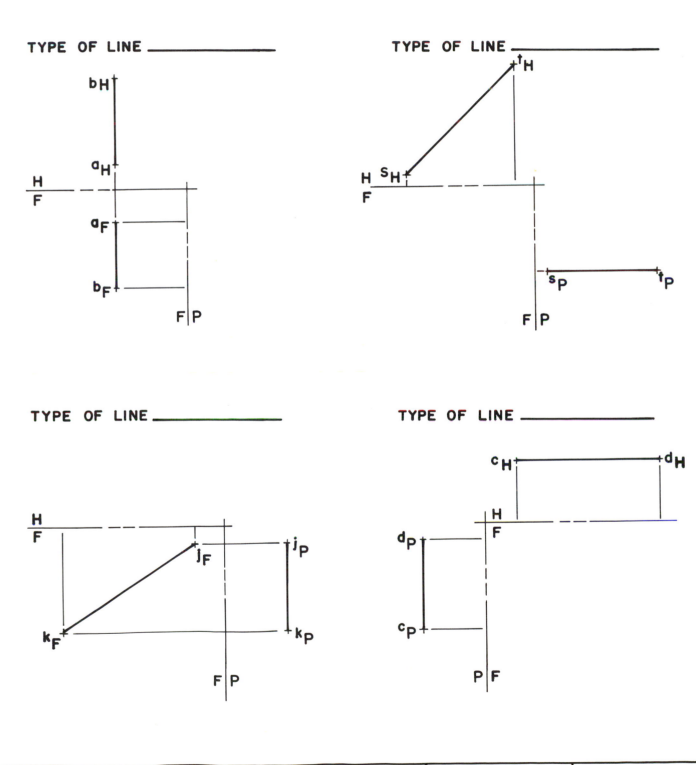

| DR. BY: | COURSE & SEC: | SCALE: | DATE: |

Problem 5: Draw the horizontal, front, and right profile views of line **AB** described below, and answer questions 5a. and 5b.

A is to the left of **B**.
A and **B** are each 1 inch below the horizontal plane.
A is ½ inch behind the front plane and 2 inches away from the

right profile plane.
B is 1 inch behind the front plane and ½ inch away fron the right profile plane.

5a. what is true length of line **AB**? _____ 5b. What type of line is **AB**? _____

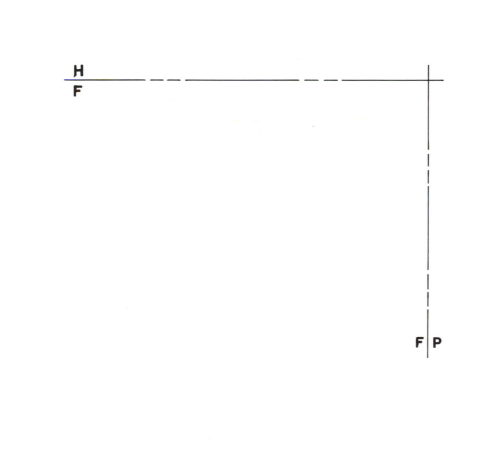

H
F

F P

Problem 6: Draw the horizontal, front, and right profile projections of line **XY** described below, and answer questions 6a. and 6b.

X is to the left of **Y**.
X and **Y** are both ¼ inch behind the front plane.
X is ⅜ inch below the horizontal plane and 2-¼ inches away

from the right profile plane.
Y is 15/16 inch away from the right profile plane.
Line **XY** makes an angle of 45° with the horizontal plane.

6a. What is the true length of line XY? _____ 6b. What type of line is XY? _____

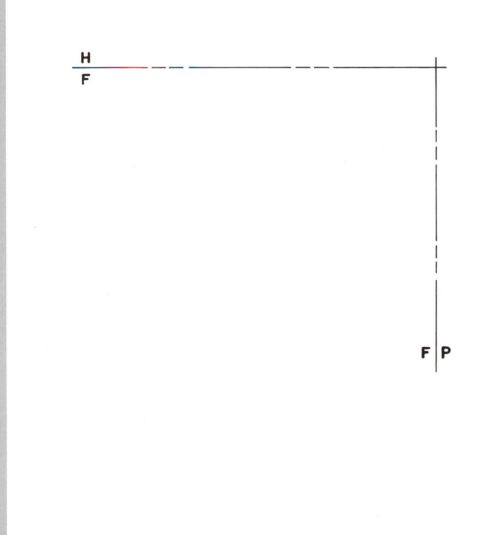

Name _____

CHAPTER TWO
TEST

1. If you were given a three view (front, top, right profile) orthographic drawing of an object, **(a)** in which view(s) would you be able to see the distance that the object lies below the horizontal plane, and **(b)** in which view(s) would you see the distance that the object lies behind the front plane, and **(c)** in which view(s) would you see the distance that the object lies to the left of the right-profile projection plane?

    (a) _____

    (b) _____

    (c) _____

2. Explain the term foreshortened *and* sketch an example that illustrates your explanation.

3. Define the following *and* sketch an example of each type of line.

    (a) Horizontal line

    (b) Frontal line

    (c) Profile line

CHAPTER

THREE

Auxiliary Views

3.1 INTRODUCTION

You have learned that it is possible for an observer to move to different positions to view an object. In addition, you have seen how to draw the different views in their proper relationship on a sheet of paper. Sometimes it is necessary to view objects or situations from various angles. Views projected on any projection planes other than the primary, or principal, planes are auxiliary views. A **primary auxiliary view** is found by projection on a plane that is adjacent and thus perpendicular to one of the six principal planes of the orthographic box. A **secondary auxiliary view** is found by projection on a plane that is adjacent and thus perpendicular to a primary auxiliary view.

Most industrial drawings require dimensions. Almost all objects drawn are boundaried by lines and planes. However, lines or planes cannot be dimensioned in a view unless they appear true size in that particular view. No piece of steel rod can be dimensioned in any view unless it is drawn true length. No piece of steel plate can be dimensioned unless it is shown true size and shape. Therefore, you must know how to find the necessary view of the object which will allow you to dimension it properly. Finding the needed view often requires the use of successive auxiliary views.

3.2 FUNDAMENTAL VIEWS

Almost all objects or situations a drafter may have to draw may be drawn through the use of one or more of four fundamental views. A thorough working knowledge of the four fundamental views will be the basis for solving all problems using descriptive geometry.

The four fundamental views are:

- The true length of a straight line.
- A straight line as a point.
- A plane as an edge.
- A plane in its true size and shape.

47

Chapter 3 will cover the first two fundamental views concerning straight lines, and upon completion of Chapter 4, you should have an understanding of the fundamental views of a plane.

3.3 TRUE LENGTH OF A LINE

A line must be parallel to the projection plane for a view to show its true length. Each of the principal line types discussed in Chapter 2 were, by definition, parallel to one of the primary planes of projection. An **oblique line** is a straight line which is not parallel to any of the six planes of the orthographic box. When a line is oblique, it will not show true length in any of the primary views. In each of these views, the line will be foreshortened. Finding the true length of the line requires the use of an auxiliary projection plane. **A line will appear true length on a projection plane that is parallel to the line**.

Figure 3-1 illustrates finding the true length of a piece of steel rod that is foreshortened in the top and front views. Auxiliary plane 1 has been drawn parallel to the rod. Therefore the auxiliary view of the rod appears true length. **The adjacent view of the line must appear parallel to the fold line between the views**. In this case, the top view of the rod (a_H, b_H) is parallel to fold line H/1. (The fold lines between views other than the primary views are labeled with consecutive numbers.) Notice where distances **X** and **Y** were found, and how they were transferred to the auxiliary view. Any point on a line or object will appear the same distance below the fold line in all elevation views that are related to the top view. You should also note that this auxiliary view is an **auxiliary elevation view**, since it is perpendicular to the horizontal projection plane, and all lines of sight for this view are horizontal (or level).

With an inclined auxiliary view, however, the lines of sight are neither vertical nor horizontal (level). Figure 3-2 shows a hip rafter of a roof in its true length on an inclined auxiliary projection plane. The auxiliary plane is perpendicular to the front projection plane; and the lines of sight are inclined. Again take note of distances **X** and **Y** and how they are transferred to the auxiliary view. **In all views** (top and auxiliary, in this case) **that relate to a common view** (the front, in Figure 3-2), **the object, or any point on the object, is the same distance away from the fold line**. In Figures 3-1 and 3-2, primary auxiliary views have been used to find the true length of oblique lines.

Let's try a sample problem. Complete Figure 3-3 to find the true length of line **CD**. First use an auxiliary elevation view (1) and then use an inclined auxiliary view (2). Views 1 and 2 are both primary auxiliary views, since each is adjacent to a primary view.

The process for finding the true length of line **CD** in auxiliary elevation view 1 is as follows:

1. The first step is to draw a fold line H/1 parallel to $c_H d_H$. This has been done for you in Figure 3-3.
2. Draw projection lines perpendicular to fold line H/1 from c_H and d_H and extending into view 1.
3. Take measurements from fold line H/F to c_F and to d_F, and transfer these to view 1.
4. Connect c_1 to d_1, and measure it to find the true length of **CD** in auxiliary elevation view 1.

To find the true length of line **CD** in the inclined auxiliary view 2, proceed as follows:

1. First a fold line, F/2, is drawn parallel to $c_F d_F$. This has been drawn for you in Figure 3-3.
2. Draw projection lines perpendicular to fold line F/2 from $c_F d_F$ and extending into view 2.
3. Take measurements from fold line H/F to c_H and d_H amd transfer these to view 2.
4. Connect c_2 to d_2, and measure it to find the true length of line **CD** in the inclined auxiliary view 2. **Lines $c_1 d_1$ and $c_2 d_2$ should be equal in length**.

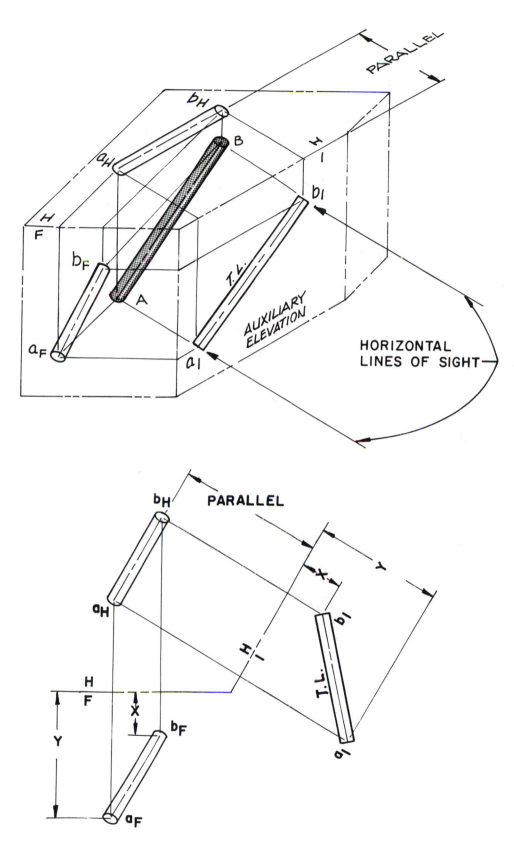

Figure 3-1. Auxiliary elevation view.

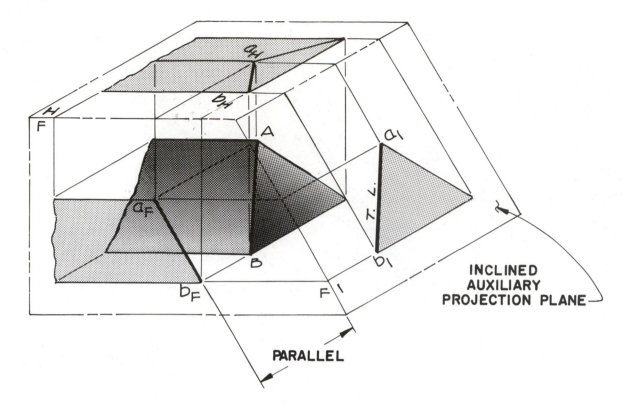

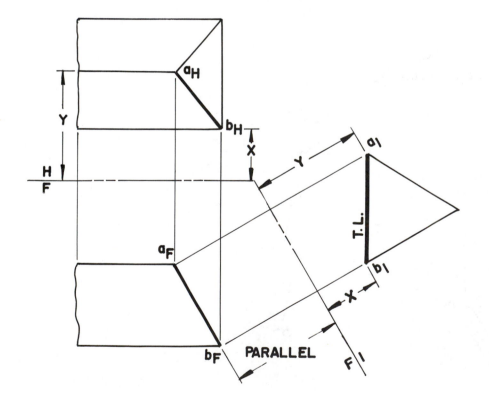

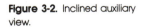

Figure 3-2. Inclined auxiliary view.

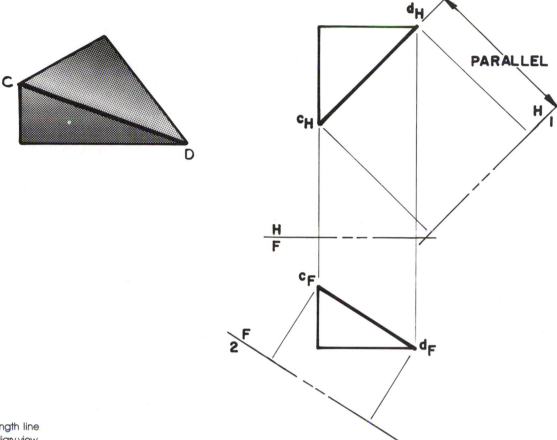

Figure 3-3. True length line using a primary auxiliary view.

3.4 BEARING OF A LINE

The bearing of a line is the angular relationship of the horizontal projection of that line relative to compass points and expressed in degrees. North is assumed to be directed toward the top of the page, unless otherwise specified. An angle less than 90° is usually given for bearing, and is given from north or south. It is customary to begin at north or south and swing to the angle in the direction indicated. Figure 3-4 is an illustration of the contour lines on a map in the horizontal view. Lines a_Hb_H, a_Hc_H, a_Hd_H, and a_He_H are drawn with various bearings. The bearing of line a_Hb_H is N 60°E, which means that the top view of line **AB** is turned away from north 60° toward the east. Likewise, line a_Hc_H has a bearing of S 30°E, meaning that the top view of line **AC** is turned away from south 30° toward the east. Line a_Hd_H is turned away from south 45° toward the west. Finally, line a_He_H is swung away from north 75° to the west. Since a compass is presumed to be held in a horizontal position, the bearing of a line is fixed regardless of whether the line is level, or it has a steep slope. For this reason the front view of the lines shown in Figure 3-4 has been deleted.

Figure 3-5 is an illustration of a piece of steel angle, **XY**. The bearing of the steel angle is N40°E, which means point y_H is turned away from north 40° toward the east. Notice that two positions of the angle are shown in the front view (x_Fy_F and $x_{F1}y_{F1}$). This is to demonstrate that bearing is not affected by slope.

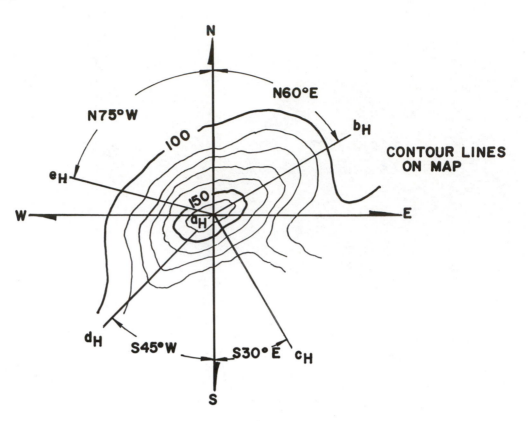

Figure 3-4. Bearing of lines.

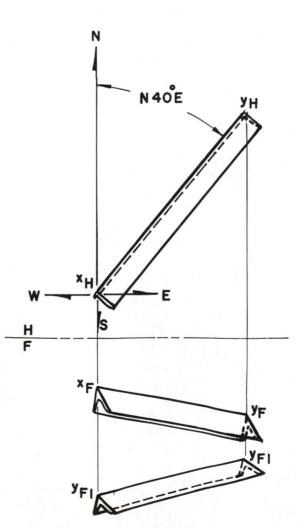

Figure 3-5. Bearing of a line.

3.5 TRUE SLOPE OF A LINE

The bearing of a line establishes its direction in the top view. **The slope of a line is the angle in degrees that the line makes with a horizontal (level) plane.** In the special case of a frontal line, the true slope is seen in the front view. But in the case of all oblique lines, **the true slope of the line can be seen only in an elevation view that shows it true length.** Remember that elevation views have horizontal (level) lines of sight. An inclined auxiliary view may show true length, but it **never** shows the true slope of a line because it does not have level lines of sight.

Figure 3-6 illustrates finding the true slope angle of a wall brace, **XY**. A fold line H/1 is drawn parallel to $x_H y_H$ to obtain the auxiliary elevation of line **XY**. This view

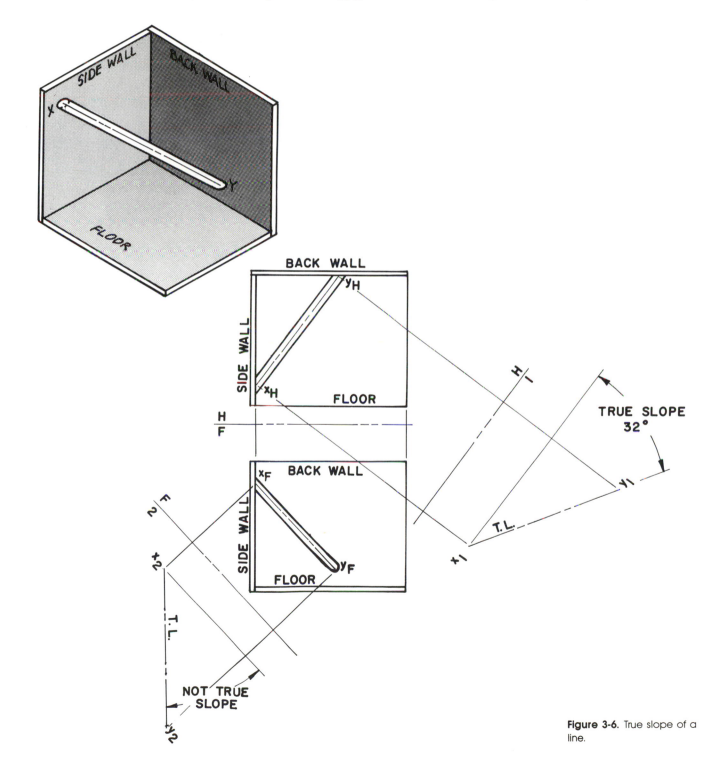

Figure 3-6. True slope of a line.

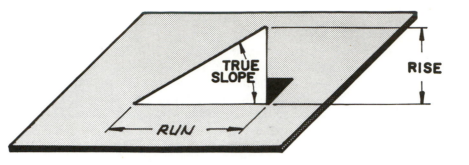

Figure 3-7. Grade of a line in relation to slope.

contains the true length and the true slope of the wall brace, **XY**. Notice that inclined auxiliary view 2 has also been drawn. This view also shows the true length of line **XY**, but it does **not** show the true slope because it is not an elevation view, having level lines of sight.

3.5.1 Grade of a Line

The **grade** of a line is another way to describe the inclination of a line in relation to a horizontal plane. The percent grade is found by evaluating the following expression: Percent grade = Rise/Run × 100. This is illustrated in Figure 3-7. Observe also that the grade is the tangent of the slope angle multiplied by 100.

Perhaps the most common method of measuring run and rise is to use an engineer's scale with divisions in multiples of ten. Figure 3-8 shows percent grade determined for a highway section. Since grade is simply another way to express slope, the conditions for finding grade are the same as for finding true slope of a line. You must have the true length of the line in an elevation view.

Figure 3-9 illustrates finding the grade of an oblique line, in this case, a guy wire, **AB**, for a radio transmission tower. The true length of wire **AB** is found in auxiliary elevation view 1. It is in this view that the true length and true slope of the guy wire is seen, and that the rise and run is measured, and that the grade is determined.

3.5.2 Other Means of Specifying Inclination

In certain engineering fields, inclination is specified differently. Each of the methods to follow are other ways of expressing the relationship of a line to a horizontal plane.

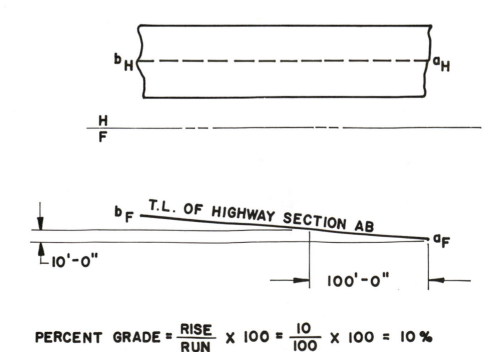

Figure 3-8. Grade of a highway.

$$\text{PERCENT GRADE} = \frac{\text{RISE}}{\text{RUN}} \times 100 = \frac{10}{100} \times 100 = 10\%$$

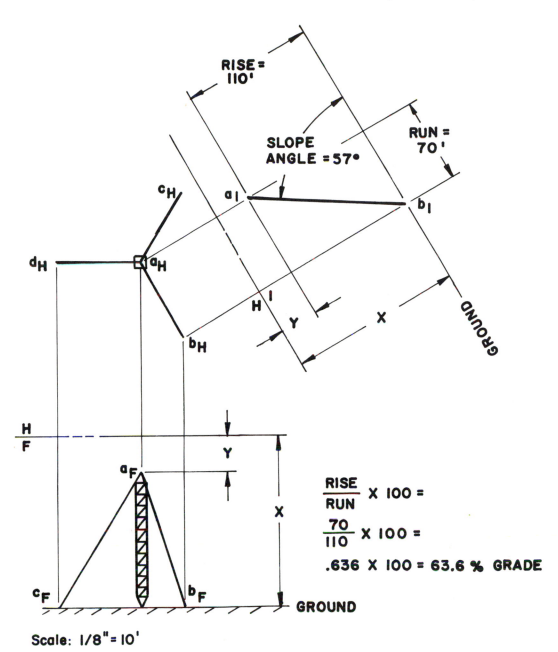

Figure 3-9. Grade of an oblique line.

In architectural drafting, slope can be seen as the **pitch** of a roof, expressed as the ratio of vertical rise to 12 feet of span. The graphical method of indicating pitch on a drawing is illustrated in Figure 3-10. In structural engineering, slope is the **bevel** of a beam and is indicated as shown in Figure 3-11. Finally, in civil engineering, **grade** is frequently used but in addition to grade, **slope** and **batter** are used. Figure 3-12 illustrates slope as it is used for an earth dam. Figure 3-13 shows the use of batter on a concrete footing.

3.6 SECONDARY AUXILIARY VIEWS

The discussion and illustration of auxiliary views has thus far been confined to primary auxiliary views. Remember these are the auxiliary views adjacent to any of the primary views. If you review the examples in this chapter, you will see that each auxiliary view used was either adjacent to the front view or to the top view of the line.

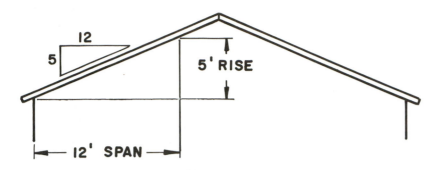

5:12 PITCH

Figure 3-10. Pitch of a roof.

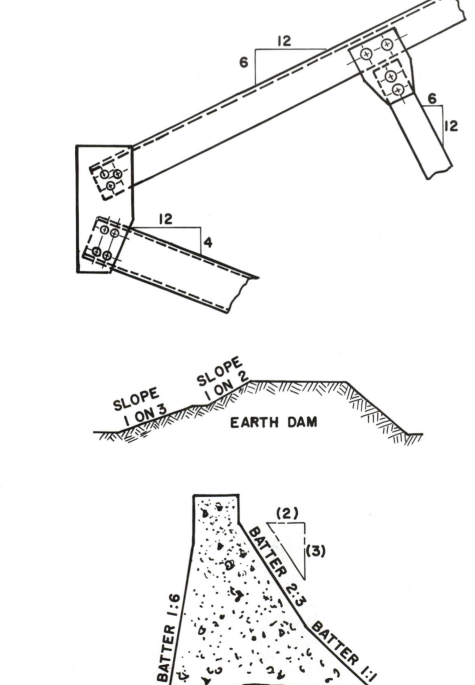

Figure 3-11. Slope (bevel) of a beam.

Figure 3-12. Slope in civil drafting.

Figure 3-13. Batter.

There are many occasions when the required solution is not found in either the primary views or in the primary auxiliary views. In these cases, it is necessary to use an **additional view that is adjacent to a primary auxiliary view and called a secondary auxiliary view**.

3.6.1 Point View of a Line

Thus far, we have used the auxiliary view to show the true length of a line (the first fundamental view). The second fundamental view is to show a straight line as a point. To do so often requires the use of a secondary auxiliary view. **To find a point view of a line, the line of sight must be parallel to the true length view of the line**. This means you must look at the end of the line. Remember, a line must appear true length in some view before a view may be drawn viewing the line as a point. Figure 3-14 shows a section of pipe, **AB**, seen as a point in secondary auxiliary view 2.

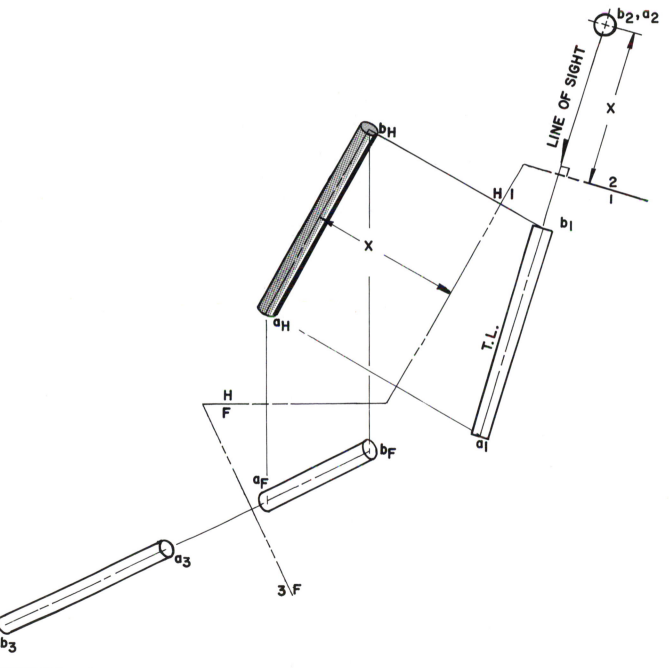

Figure 3-14. Secondary auxiliary view: Point view of a line.

The procedure for finding the point view of pipe **AB** is as follows:

1. Draw fold line H/1 parallel to $a_H b_H$ to obtain the true length view $a_1 b_1$.
2. In order to have lines of sight parallel to the true length view of line **AB**, fold line 1/2 must be drawn perpendicular to the true length view, $a_1 b_1$. (Remember that auxiliary views are labeled with consecutive numbers.)
3. A projection line is drawn perpendicular to fold line 1/2 from $a_1 b_1$, and distance **X** is transferred to auxiliary view 2, where the end view of pipe **AB** is seen.
4. Primary auxiliary view 3 is drawn for illustrative purposes to show that a point view will **not** result, if the line is not drawn in true length first.

CHAPTER PROBLEMS

The problems in this chapter are based on a variety of industrial applications. In solving these problems you should try to utilize direct approaches that get to the heart of the problem. At this analysis stage of the design process, many problems use centerlines and single lines to represent the problem situation. Yet, when necessary, relevant sizes are given.

Problem 1: A framework made of square steel tubing welded together is shown using single lines to represent the tubing. It is necessary to check the alignment of the finished form by checking the measurements of diagonal distances **AB**, **CD**, and **AE**. Find the true length of each diagonal. Scale: $\frac{1}{2}" = 1'\text{-}0"$

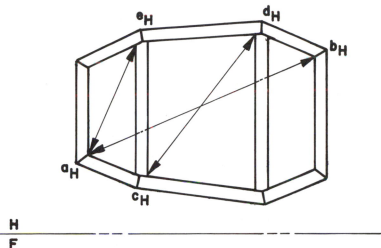

H
F

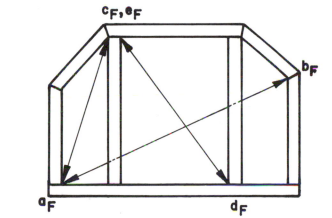

| DR. BY: | COURSE & SEC: | SCALE: | DATE: |

Problem 2: A section of concrete wall with a level top is shown. Corners **AB** and **CD** are to be covered with steel angle for protection against chipping. Find the true lengths and true slope angles of corners **AB** and **CD**. Scale: ¼" = 1'-0"

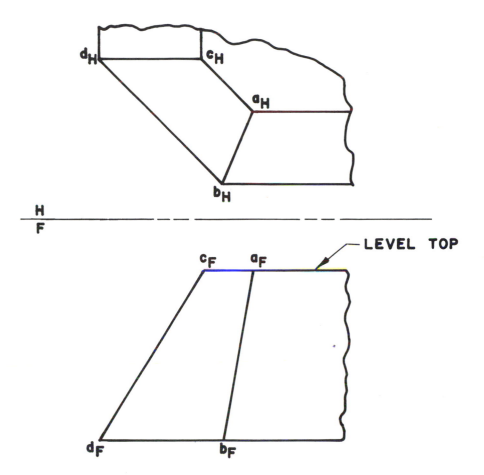

Problem 3: The front and side views of the end panel of a bridge are shown. Find the true length and the true slope angle of diagonal member **XY**. Scale: $1'' = 10'$

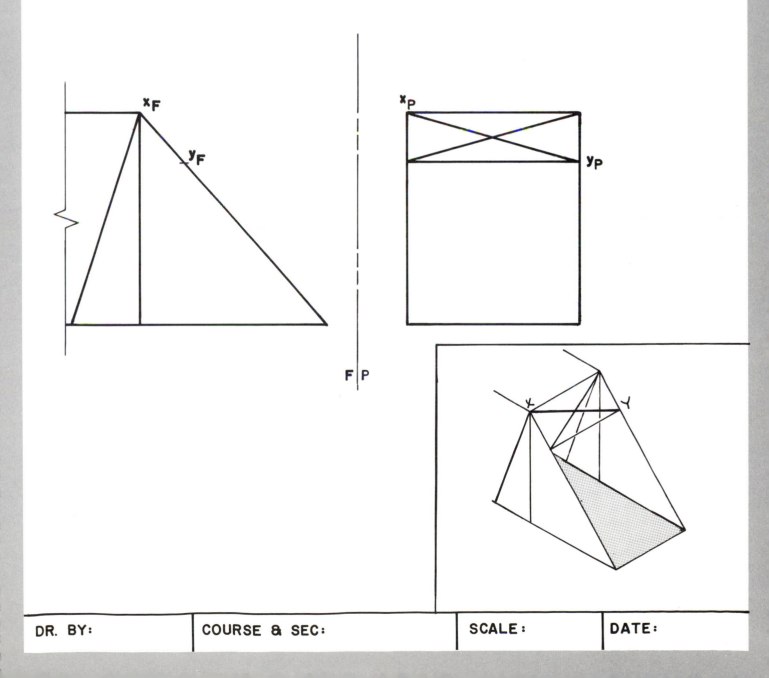

X_F

y_F

X_P

y_P

F | P

| DR. BY: | COURSE & SEC: | SCALE: | DATE: |

Problem 4: A derrick, supporting a horizontal derrick boom, YZ, is mounted on a barge as shown. Find the true lengths and the true slope angles of the anchor cable, **WY**, and the structural member, **XY**. Scale: 1" = 20'

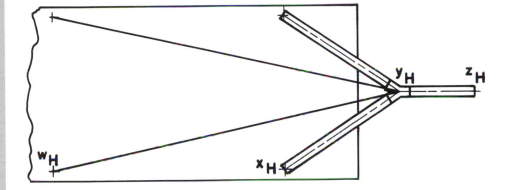

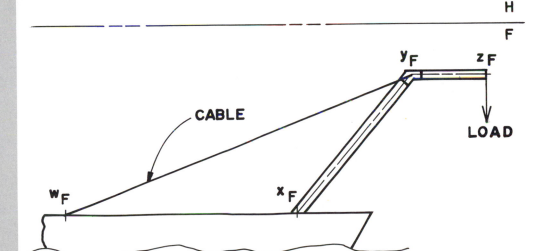

CABLE

LOAD

Problem 5: This drawing and the accompanying information were brought to you by a survey party. Determine the elevations of towers **C** and **D**, and the true distance between them.

The elevation of **A** is 320 feet above sea level, while the elevation of **B** is 280 feet. The slope angles from **A** to **C** and from **B** to **D** are both 55°. Scale: 1" = 100'

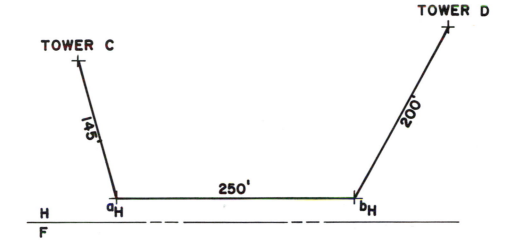

TOWER D

TOWER C

145'

200'

250'

H
F

a_H

b_H

(EL. 320') a_F

b_F (EL. 280')

Problem 6: Two pipes, **AB** and **CD**, that intersect at **D**, both slope downward in the direction of the arrows, as shown. Both pipes have the same slope, and **AB** is actually 130 feet long.

Find the difference in elevation between **A** and **C**, and the true length of **CD**. Show both pipes in the front view. Scale: 1" = 40'

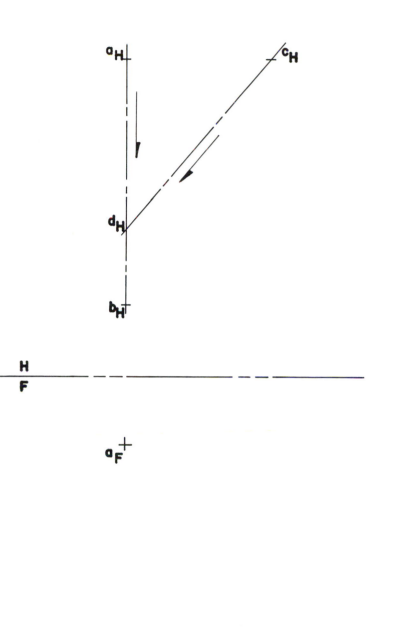

Problem 7: The property line of a building site is shown. Find the bearing of each segment of the property line.

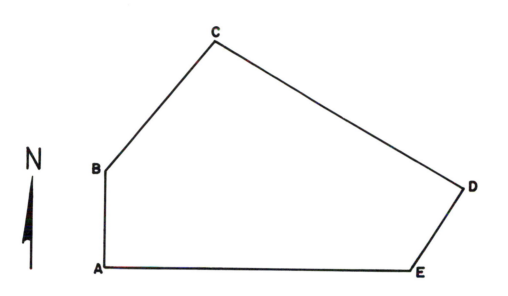

| DR. BY: | COURSE & SEC: | SCALE: | DATE: |

Problem 8: Draw two views of a straight highway section, **AB**, that is 250 feet long, and runs N47°W with a downgrade of 6 percent. Scale: 1" = 100'

$+$ a$_H$

H
F

$+$ a$_F$

| DR. BY: | COURSE & SEC: | SCALE: | DATE: |

Problem 9: Draw two views of a mine shaft, **MN**, beginning at the bottom of the mine shaft (point **M**), and bearing S35°W at an uphill grade of 15 percent for 150 feet. Scale 1" = 100'

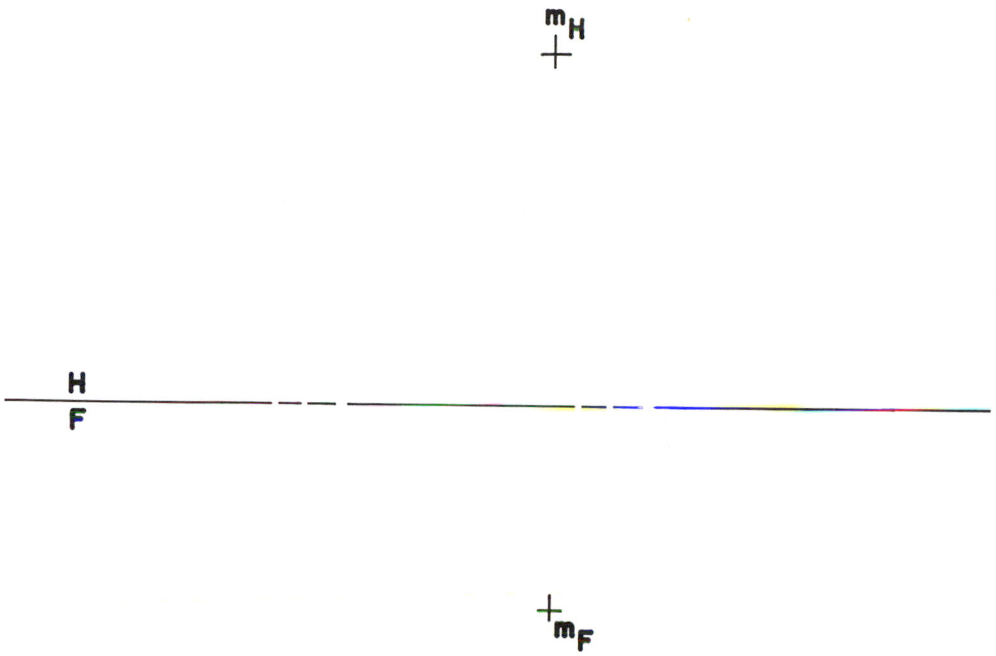

m H
+

H
F

+
m F

DR. BY:	COURSE & SEC:	SCALE:	DATE:

Problem 10: A conveyor line system is currently installed as indicated by the following information.

SECTION	BEARING	SLOPE/GRADE	TRUE LENGTH
AB	N40°W	−20 percent	135 feet
BC	Due West	0°	105 feet
CD	S30°W	−15°	150 feet

A revision in production sequence calls for the elimination of the present conveyor system and the installation of a new conveyor line directly from **A** to **D**. Find the true length, bearing and slope of the new line. The conveyor may be represented as single lines for your solution. The front and top views should include both the new and the old systems. Scale: 1″ = 100′

$+ a_H$

H
F

$+ a_F$

Problem 11: A vertical mast, **XY**, is anchored by three guy wires, **XA**, **XB**, and **XC**. Anchor **A** is 13 feet west and 4 feet north of mast **XY** and is at an elevation of 140 feet. Anchor **B** is 12 feet east and 7 feet north of mast **XY**, and is at an elevation of 146 feet. Anchor **C** is 5 feet east and 10 feet south of the mast and is at an elevation of 138 feet. A horizontal boom, pivoted at **Z**, is to swing clear around the mast and must clear all guy wires. Find the true length of each guy wire, and the length of the longest horizontal boom that can be used. Show the guy wires and the path of the end of the horizontal boom in the front and top view. Scale ⅛" = 1'-0"

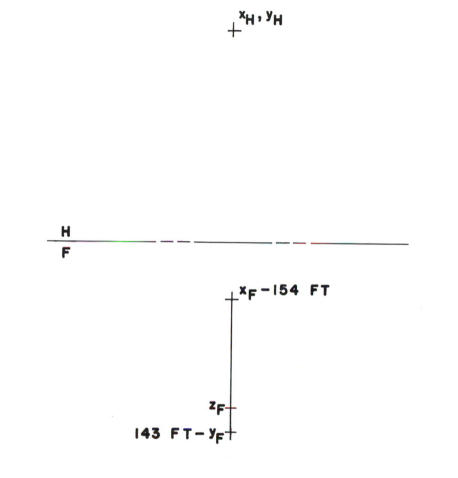

x_H, y_H

H
F

$x_F - 154$ FT

z_F

143 FT $- y_F$

Problem 12: From point **A** an old sewer pipe runs 12 feet at N41°E on a 10 percent downgrade. From point **C**, which is 15 feet east and 6.5 feet north, and 24 inches lower than point **A**, a new sewer, 13 feet long is planned. From point **C** it will run N54°W at a 18 percent downhill grade. Both pipes are 12 inches in diameter. Will the new pipe pass under or over the old one, and what will be the vertical clearance (if any) where they cross? Trace the given information and complete your solution on a B-size drawing sheet.) Scale: ¼" = 1'-0"

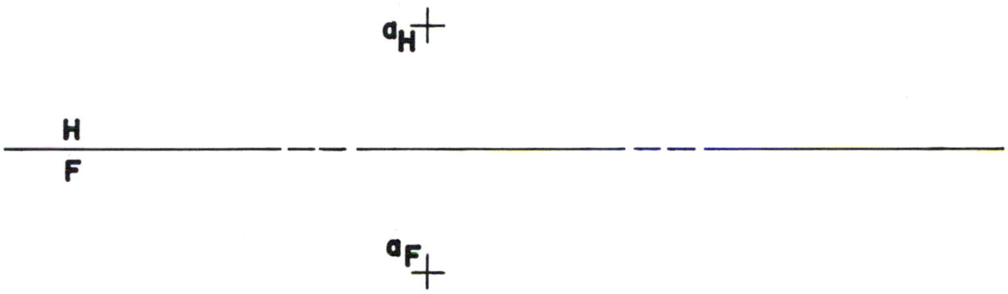

a_H +

H
—————————————————————————
F

a_F +

| DR. BY: | COURSE & SEC: | SCALE: | DATE: |

Problem 13: A steel frame used for a hoist is shown. What are the true lengths of structural members **BD** and **AF**? What is the angle between side braces **BD** and **BE**? (Hint: A point view of **AC** will help with this.) Scale: ¼" = 1'-0"

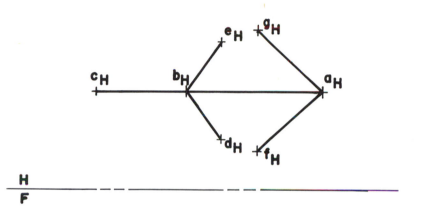

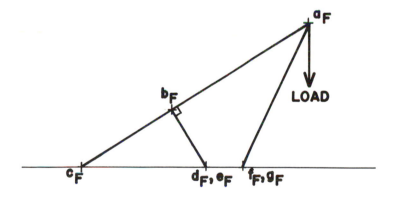

Problem 14: Spur gears mesh correctly with each other when their pitch diameters are just in contact (tangent). The center line, **WX**, of a shaft for a 3-inch pitch diameter gear is shown. Only the top view of the centerline, **YZ**, of a shaft carrying an 8-inch pitch diameter gear is shown. Locate the front view of shaft **YZ**, so that the gears will mesh properly. Choose the position of **YZ** that is lower than shaft **WX**. Determine how much lower **Y** is than **W**. Show a view that will allow you to see the pitch diameters as tangent circles. (Note: The shafts are parallel.) Scale: ¼" = 1"

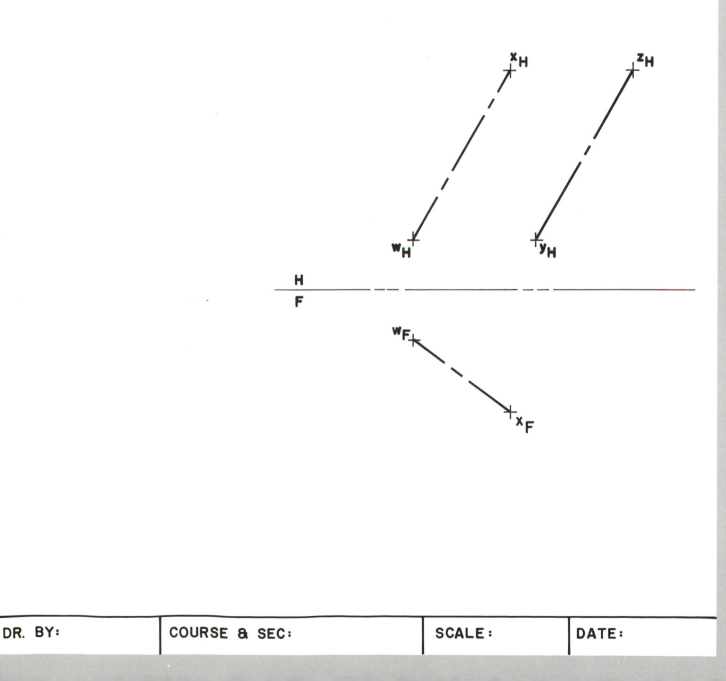

CHAPTER THREE
TEST

1. What is an auxiliary view?

2. What is the purpose of an auxiliary view?

3. Describe the difference between a primary and secondary auxiliary view.

4. List the two fundamental views covered in Chapter 3.

5. What is an oblique line?

6. Describe how to find the true length of a line. Include an explanation of how distances are transferred to obtain the location of the true length view, and why you used the distances you chose.

7. Draw an example of your answer to question 6.

8. Indicate whether the following statements are true or false. Provide a written explanation justifying your answer.

    T F   a. Two successive projection planes are always at right angles to each other.

    T F   b. In order to obtain a point view of a line, your line of sight may be selected parallel to any view of the line.

9. Explain the difference between an auxiliary elevation view and an inclined auxiliary view.

10. Define the bearing of a line.

11. If a line has a bearing of S 60°W, what does this mean?

12. Explain the relationship between bearing and slope.

13. Define true slope of a line.

14. Use a sketch to illustrate how you would find true slope.

15. Define grade of a line.

16. Determine the grade of the line used in your sketch for question 14. (Show your work.)

17. List three ways, other than true slope angle and grade, to express the relationship of a line to a horizontal plane.

18. What two conditions must exist before you can measure true slope or determine grade?

19. What two conditions are necessary to view a line as a point?

CHAPTER

FOUR

Planes

4.1 INTRODUCTION

Almost all objects and many engineering problems consist of plane surfaces. The basic principles involving planes are applicable in most industrial fields. **A plane is a surface which is not curved or warped. It is a surface in which any two points may be connected by a straight line, and the straight line will always lie completely within the surface**. In other words, every point on that line is also on the surface of the plane. Figure 4-1 is a pictorial drawing of an object composed of series of planes. Surfaces **A**, **B**, **C**, and **D** each satisfy the definition of a plane. You could place a pencil on any one of these plane surfaces, in any position, and it would lie flat on the surface of the plane.

So far the various projection planes (horizontal, front, profile, and auxiliary) have been discussed in relation to lines. This chapter deals with the representation of plane surfaces in space and their projections onto the standard projection planes.

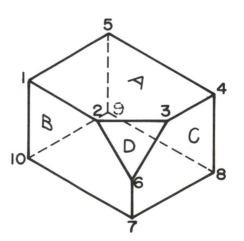

Figure 4-1. Pictorial drawing: Planes.

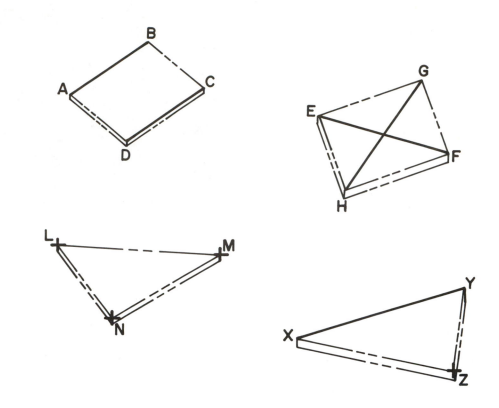

Figure 4-2. Formation of a plane: Two parallel lines.

Figure 4-3. Formation of a plane: Two intersecting lines.

Figure 4-4. Formation of a plane: Three points.

Figure 4-5. Formation of a plane: A line and a point.

4.2 THE FORMATION OF PLANES

A plane may be constructed in four different ways, each of which is commonly used in engineering work. Figures 4-2, 4-3, 4-4, and 4-5 show the four methods of representing a plane.

Two parallel lines form plane **ABCD** in Figure 4-2. In conjunction with this figure, examine Figure 4-1, and notice that parallel lines **1-2** and **7-10** could be said to form plane **B**. Two intersecting lines, **EF** and **GH**, form plane **EHFG** in Figure 4-3. Any three points, not in a straight line, form plane **LMN** in Figure 4-4. Points **2, 3,** and **6** in Figure 4-1 form plane **D** in the same manner. A point, **Z**, and a line, **XY**, in Figure 4-5 form plane **XYZ**. Plane **D** in Figure 4-1 could also be a plane formed by line **2-3** and point **6**.

4.3 PLANES: NORMAL AND OBLIQUE

A plane surface that is parallel to any one of the primary projection planes, such as the top or front planes, is a **normal plane**. Whereas a plane which is inclined to all primary projection planes is called an **oblique plane**. Figure 4-6 shows a pictorial and an orthographic drawing of an object showing all six primary views. Examine this object carefully.

Planes **A, G, E,** and **F** are normal because they are each parallel to at least one of the primary planes. For example, imagine yourself looking at the front of this object. In the front view, all you see of plane **A** is its edge. It looks like a straight line. That edge view of plane **A** is parallel to fold line H/F which means the plane is parallel to the horizontal projection plane. Because plane **A** is parallel to the horizontal projection plane, its true size and shape can be seen in the horizontal view. Ask yourself in which view you see plane **E** as an edge view, and in which view you see its true size and shape. Do the same with planes **G** and **F**.

Planes **B, C,** and **D** are oblique because they are each inclined to each of the primary planes. Because of their position, the true shape and size of these planes is not seen on any of the primary planes.

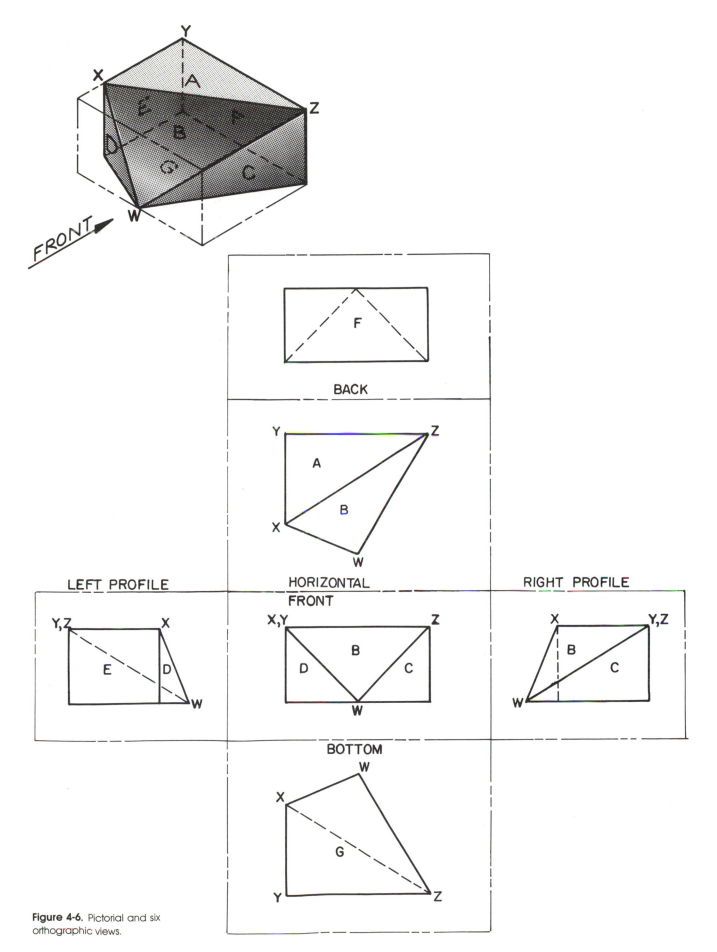

Figure 4-6. Pictorial and six orthographic views.

4.4 EDGE VIEW OF AN OBLIQUE PLANE

Any plane will be seen as an edge, that is, will appear as a straight line, in a view in which any line on the plane appears as a point. This is the third fundamental view. You have already learned how to view a line as a point. **Remember, the point view of a line is found perpendicular to the view in which it appears true length.**

This concept is used to find the edge view of a plane. Let's return for a moment to Figure 4-6. We have already established that plane **A** is seen true shape and size in the top view. Therefore, all lines in plane **A**, lines **XY**, **YZ**, and **ZX**, are true length. Fold line H/F is perpendicular to line $x_H y_H$, and line **XY** is seen as point $x_F y_F$ in the front view. This causes plane **A** to appear as an edge in the front view. Identify lines that appear in true length in plane **E**, and find the point views and thus the edge view of plane **E**.

Now let's examine the same object as is used in Figure 4-6 and study plane **B** only. In Figure 4-7 plane **B** is identified and each of three corners of plane **B** are identified (**X,Z,W**). Line $x_F z_F$ is parallel to fold line H/F, therefore it is true length in the horizontal view ($x_H z_H$). Fold line H/1 is drawn perpendicular to the true length line, $x_H z_H$, and the line of sight will be parallel to the true length of **XZ**. In view 1 line **XZ** appears as a point (z_1, x_1), causing plane **B** to appear as an edge.

Figure 4-8 illustrates an object very similar to the one shown in Figure 4-6. Only in this example plane **A** is inclined, and therefore none of the lines in plane **B** are true length in the given views. Figure 4-9 demonstrates how to find the edge view of plane **B**, now that no true length lines are available. **The solution is to create a true length line in the plane.**

If you draw a line from z_F to line $x_F w_F$, so that it is parallel to fold line H/F, this line touches $x_F w_F$ at p_F. Now project point **P** into the horizontal view. Since point **P** is on line **XW** in the front view, it will project to the top view as point p_H on line $x_H w_H$. When you connect point z_H to point p_H, you have a true length line (a horizontal line), which you can view as a point (in view 1) to find the edge view of plane **B**. To find this edge view, draw

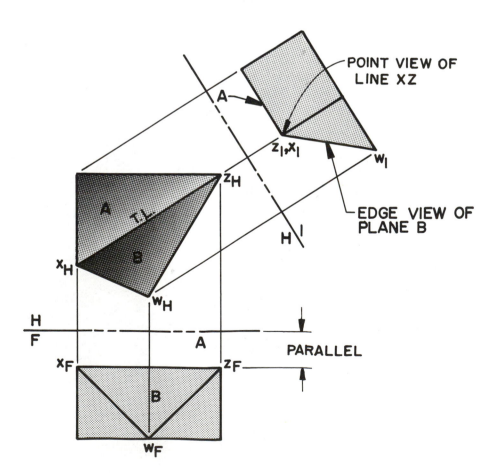

Figure 4-7. Edge view of a plane.

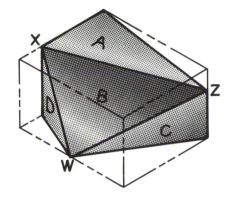

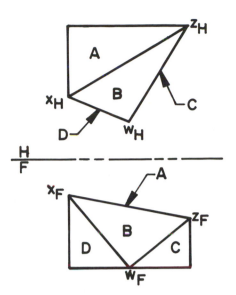

Figure 4-8. Pictorial and orthographic drawing: Plane A is inclined.

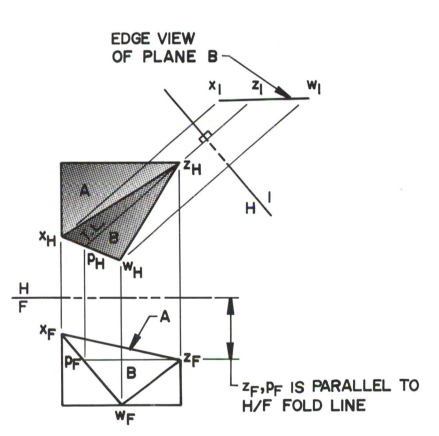

Figure 4-9. Edge view of a plane.

fold line H/1 perpendicular to line $z_H p_H$, and then project points x_H and w_H into auxiliary view 1. In auxiliary view 1 points x_1, z_1, and w_1 are connected to form the edge view of plane B. **In actuality, creating this true length line has merely moved you (the observer) into the proper location to see plane B as an edge view.**

4.5 TRUE SLOPE OF A PLANE

The true slope angle of a plane, also called the dip angle, is the angle the plane makes with a horizontal projection plane. This angle will be seen in any elevation view, where the given plane appears as an edge. In Figure 4-10, plane A is seen as

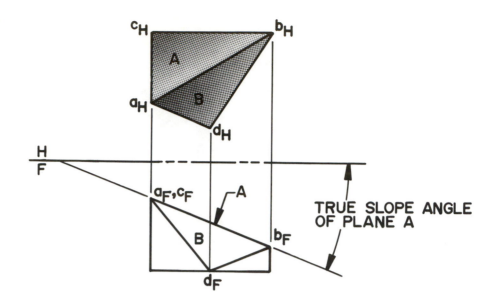

Figure 4-10. True slope of a plane.

an edge view in the front view (an elevation view), which is a view adjacent to the horizontal plane. Both conditions have been met. Here you see the true slope angle of plane **A**. But you cannot see the true slope angle of plane **B** in the given views. Why? Because plane **B** is not an edge view in the given elevation (front) view.

In order to find the true slope angle for plane **B**, you must find plane **B** as an edge in an elevation view. Return to Figure 4-9 to refresh your memory on how to find an edge view of a plane. In Figure 4-11, plane **B** is seen as an edge in view 1, and the true slope angle of plane **B** can be measured as indicated.

To allow you practice with these concepts, find the edge view and true slope of plane **A** in Figure 4-12.

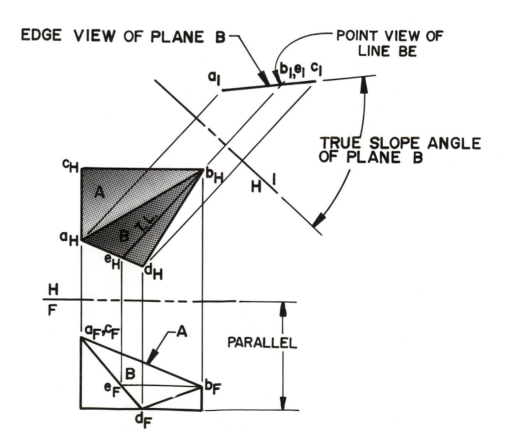

Figure 4-11. Edge view and true slope of plane B.

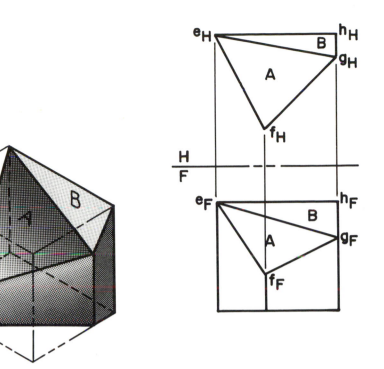

Figure 4-12. Exercise: Practice finding the edge view and true slope of a plane.

4.6 TRUE SHAPE OF A PLANE

The true or actual shape and size of any plane surface is seen on a projection plane that is parallel to the plane surface. This is the fourth and final fundamental view. Pay particular attention to the fact that each fundamental view depends on the previous fundamental view for its construction. In order to find the true shape of a plane, either normal or oblique, an edge view must first exist. In Figure 4-13 plane **A** is a normal plane. It appears as an edge view on the front projection plane. Because the edge view of the plane is parallel to the horizontal plane, indicated by the edge view of plane **A** being parallel to the H/F fold line, plane **A** appears in its true shape in the top view.

In the case of an oblique plane, it is necessary to first find the edge of the plane in question. In Figure 4-14 after finding the edge view of plane **A** in auxiliary view 1, a projection plane is placed parallel to it (fold line 1/2). The lines of sight will be perpendicular to the edge view, giving you the true shape and size of the plane in auxiliary view 2. Notice distances **X**, **Y**, and **Z** and from where they were obtained. Remember, any point on the object remains the same distance away from the fold lines in all views (the horizontal view, and view 2, in this example) that relate to a common view (view 1, in this case).

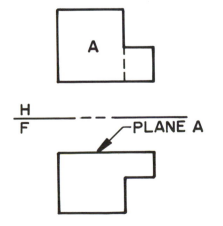

Figure 4-13. True shape of a plane.

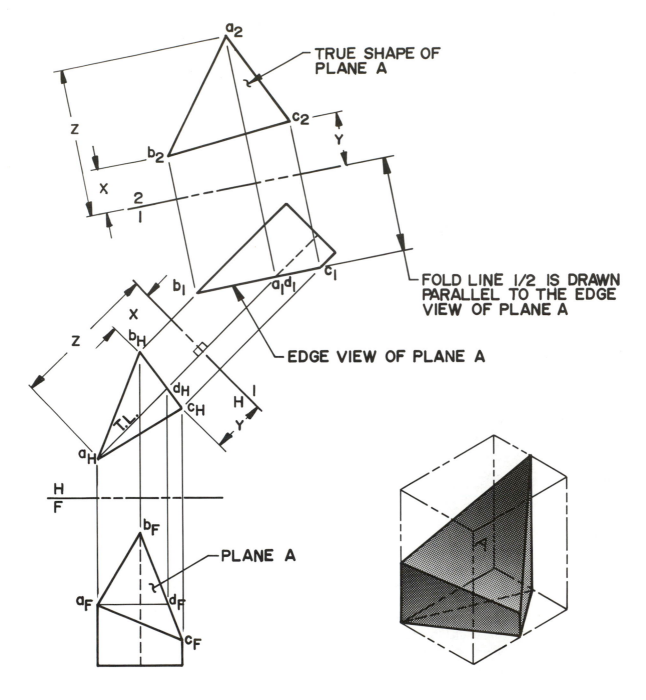

Figure 4-14. True shape of a plane.

To insure that you understand the concepts and the procedures for finding the edge view, the true slope, and the true shape of a plane, find them for plane **B** in Figure 4-15. Remember, you need to follow the steps listed below:

1. Draw a line that appears in true length in the horizontal view (a horizontal line).
2. Draw a fold line perpendicular to the true length horizontal line and find the point view of this line and, therefore, an edge view of plane **B**.
3. Draw a fold line parallel to the edge view of plane **B** in order to find the true shape of the plane.

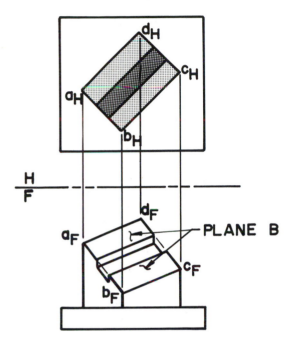

Figure 4-15. Exercise: Practice finding the edge view, true slope, and true shape of plane B.

CHAPTER PROBLEMS

Problem 1: The front and right profile views of the Jig Angle are shown. Draw a view that shows the inclined surface **A** in true shape. Fold lines should be included and labeled correctly.

(Dimensions given are true dimensions to assist you in the completion of the true shape view.) Scale: Full

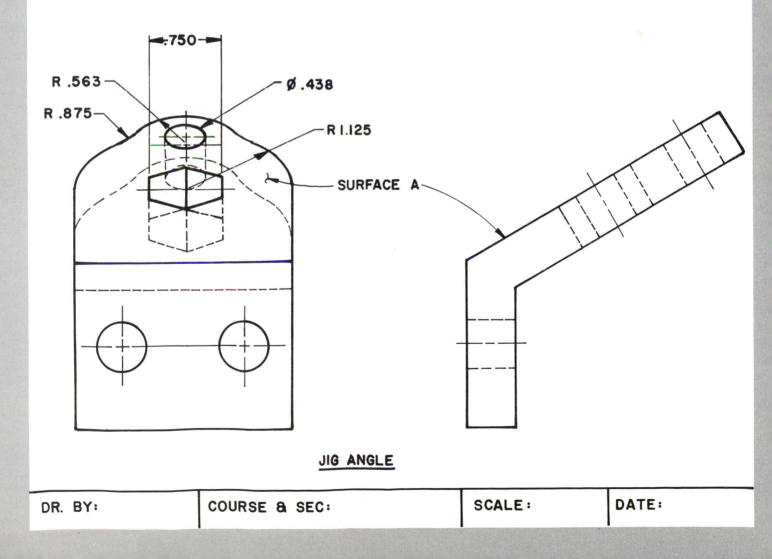

.750

R .563

R .875

Ø .438

R 1.125

SURFACE A

JIG ANGLE

| DR. BY: | COURSE & SEC: | SCALE: | DATE: |

Problem 2: The top and front views of a Dovetail Clip are shown. Find the true shape of surface **A**. For machining purposes, find the true slope of surface **A**. Fold lines should be included.

(Dimensions given are true dimensions to assist you in the completion of the true shape view.) Scale: Full

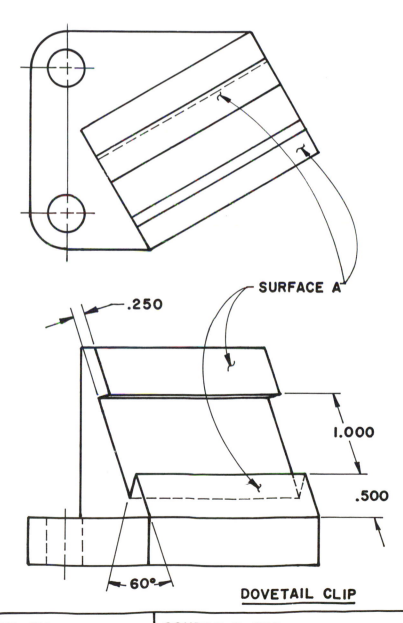

SURFACE A

.250

1.000

.500

60°

DOVETAIL CLIP

| DR. BY: | COURSE & SEC: | SCALE: | DATE: |

Problem 3: The top and front views of a Clutch Stop are shown. Find the true shape of surface **Z**. For machining purposes, find the true slope of surface **Z**. Include all fold lines. Scale: Full

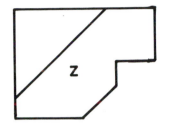

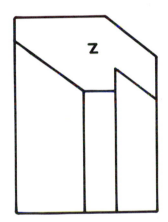

CLUTCH STOP

Problem 4: The front, top, and right profile views of a Gage Block are shown. Find the true shapes of surfaces **A**, **C**, and **D**. Include all fold lines. Scale: Full

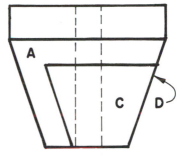

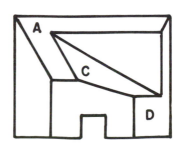

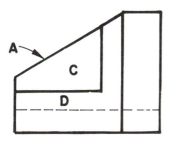

GAGE BLOCK

| DR. BY: | COURSE & SEC: | SCALE: | DATE: |

Problem 5: Two views of an A-frame structure connecting to the floor and the wall of a building are shown. Find the true size of planes **DAB**, **DAE**, and **BAC** in order that a bracket for the connection at **A** may be designed. Scale: ⅛" = 1'-0"

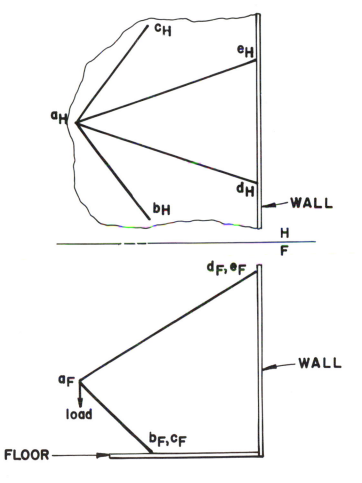

| DR. BY: | COURSE & SEC: | SCALE: | DATE: |

Problem 6: A transistion piece for a fume exhaust duct system is shown. Find the true size of planes **A, B, C,** and **D.** The interior surfaces of the duct are to receive a protective coating of epoxy resin paint. How many square feet of metal will need paint? (Disregard metal thickness.) Scale: ¼" = 1'-0"

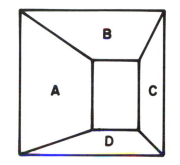

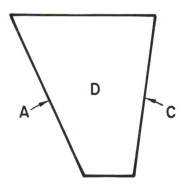

Problem 7: Two views of a flume transistion piece are shown. Plane **ABCD** is one of the side walls of this transition piece, which is to be made of reinforced concrete. Find the true shape of **ABCD** so that the forms for this piece can be made. Scale: ¼" = 1'-0"

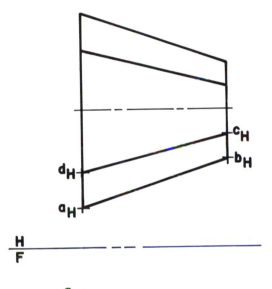

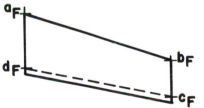

Problem 8: In the construction of pipelines it is often necessary to determine the true angle of the bend in the pipe. Find the true angle between the centerlines of pipes **AB** and **BC**. Construct a bend with a 36-inch radius between the pipe sections. Pipe sections **AB** will have a 90° connection (Tee) installed four feet from **A**. Show the position of the connection in all views. Scale: 1" = 4'

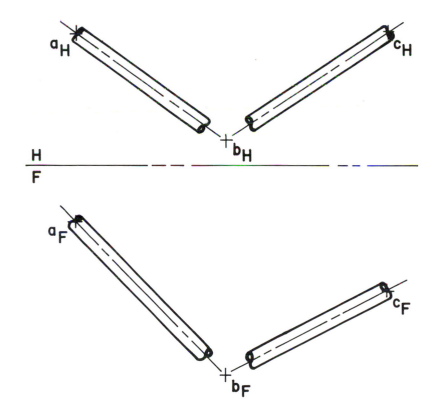

Problem 9: The top and front views of a section of rigid tubing used for oil in a gas engine are shown. It is necessary to know the true angles of the bends at **ABC** and **BCD**. Scale: ½

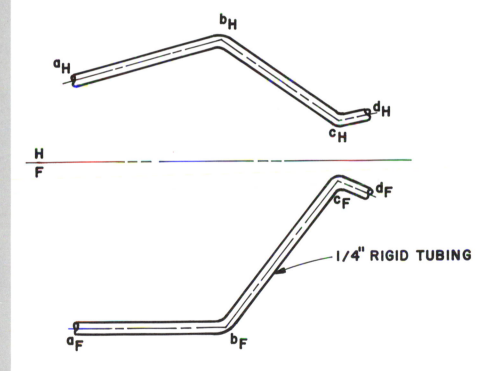

H
—
F

1/4" RIGID TUBING

| DR. BY: | COURSE & SEC: | SCALE: | DATE: |

Problem 10: Two views of the centerline of the main pipeline, **ST**, are shown. Locate a view of branch line from **R** to **ST** that connects with a standard 45° Y fitting. What is the length of the branch line? The arrowheads indicate the direction of flow. Scale: 1" = 60'

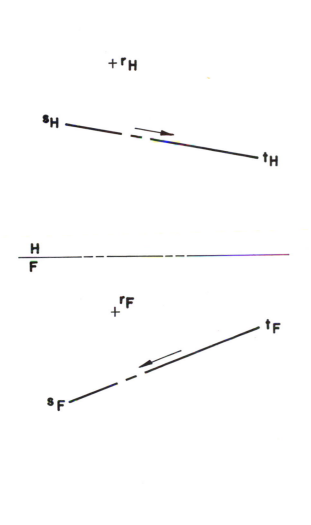

Problem 11: Three points (**A**, **B**, and **C**) on the centerline of a 24-inch conduit for conveying water to a power plant are shown. The conduit design calls for a long radius elbow to connect the straight sections of pipe. The radius is to be eight times the diameter of the pipe. Find the true size of the sweep angle of the elbow, and the true lengths of the straight pipes from **A** to the elbow and from the elbow to **C**. Show the centerline of the elbow in the top and front views. Scale: 1" = 40'

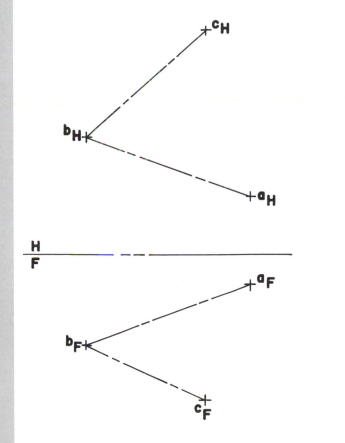

DR. BY:	COURSE & SEC:		SCALE:	DATE:

Problem 12: A 4-inch diameter pulley is oriented to carry a flat belt from **M** to **N** around a turn near **P**. Show the pulley diameter and the angle of belt contact. In addition, show the pulley diameter in the front and top views, disregarding its thickness. Scale: ¼

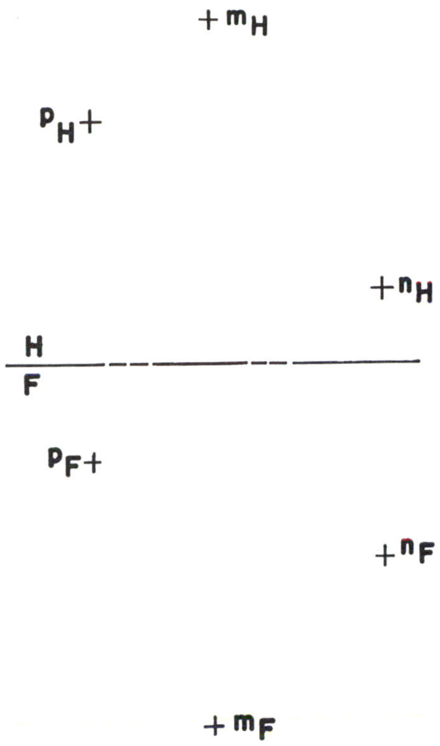

$+^mH$

$^pH+$

$+^nH$

H
───── ── ── ── ──── ─────
F

$^pF+$

$+^nF$

$+^mF$

| DR. BY: | COURSE & SEC: | SCALE : | DATE : |

CHAPTER FOUR

TEST

1. Define the term *plane* as used in descriptive geometry.

2. What is a normal plane?

3. What is an oblique plane?

4. Describe fully the procedure for finding the edge view of a plane surface.

5. Illustrate the procedure described in question 4.

6. Define the true slope of a plane.

7. What two conditions must be met in order to find the true slope of a plane?

8. Define the true shape of a plane.

9. Explain how to find the true shape of a plane.

10. Describe the relationship between the four fundamental views.

11. Indicate whether the following statements are true or false. Provide a written explanation justifying your answer.

   T F a. The true length of any line in a plane is available in the edge view of the plane.

   T F b. The true angle between two lines is seen in a view that shows either one of the lines in true length.

CHAPTER FIVE

Parallel and Perpendicular Lines

5.1 INTRODUCTION

Having studied the concepts of orthographic projection and the four fundamental views, you are well versed in the basics of descriptive geometry. The purpose of this and the following chapters is to build on those skills. You will be introduced to a variety of line and plane problems which occur often in actual practice and which drafters and engineers should know how to solve. You will continue to use the same methods of logical thinking and three-dimensional visualization you applied to the problems in previous chapters.

5.2 PARALLEL LINES

When lines are parallel, they are an equal distance from each other throughout their length. **If lines are parallel to each other, they will appear parallel in all views, except two.** Those two views are when the lines appear one behind the other, and when the lines appear as end views or points. Figure 5-1 illustrates parallel lines **AB** and **CD** on the Guide Block. Notice that they appear parallel to each other in each view.

In Figure 5-2, the two exceptions are illustrated. Two pipes, **AB** and **CD**, are hung equidistant from a ceiling. In the front and right-profile views, they appear one behind the other, while in the top view they appear parallel and in true length. In auxiliary view 1 both centerlines appear as points. Note that auxiliary view 1 also shows the true distance between the centerlines of the two pipes.

Let's go through one more example that demonstrates the principle that parallel lines will appear parallel in all views. In Figure 5-3 you are given four views of two parallel shafts. In the front and top views the shafts appear parallel. The true lengths of the shafts are shown in view 1. In view 2, the shafts are seen as points, which also allows you to see the true distance between the shaft centerlines. Try to visualize these two shafts in space as you study these four views.

On occasion it is necessary to construct a line parallel to an existing line and through a given point. Figure 5-4 illustrates this process. At (a), you are given shaft **AB** and one

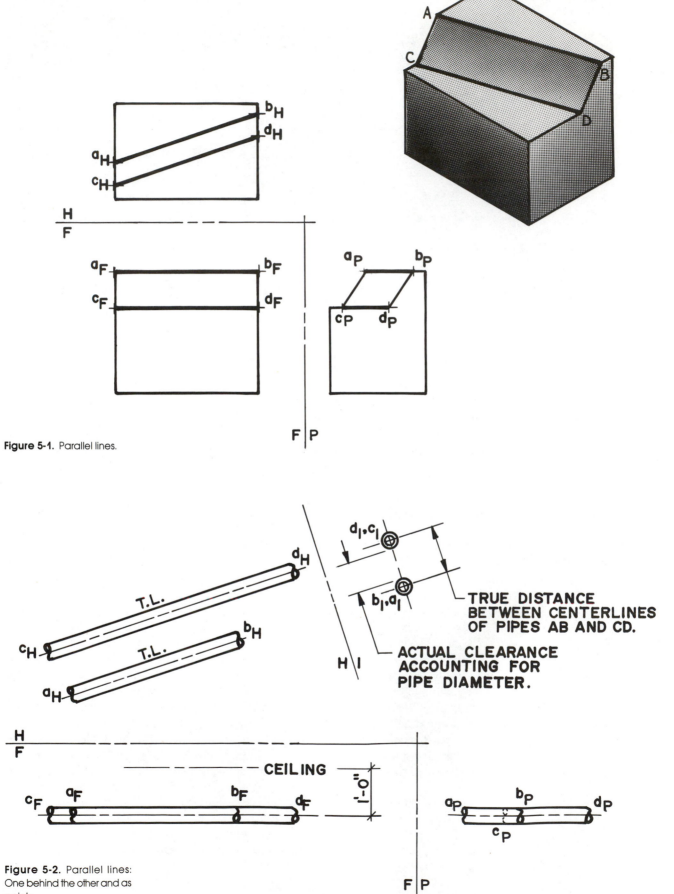

Figure 5-1. Parallel lines.

T.L.

T.L.

$d_I \cdot c_I$

$b_I \cdot a_I$

**TRUE DISTANCE
BETWEEN CENTERLINES
OF PIPES AB AND CD.**

**ACTUAL CLEARANCE
ACCOUNTING FOR
PIPE DIAMETER.**

CEILING

1'-0"

Figure 5-2. Parallel lines:
One behind the other and as
points.

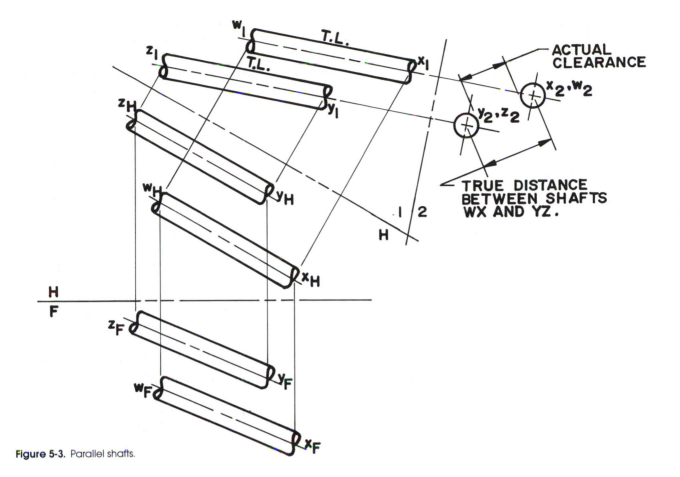

Figure 5-3. Parallel shafts.

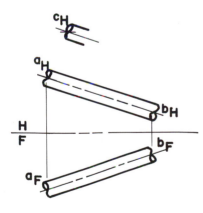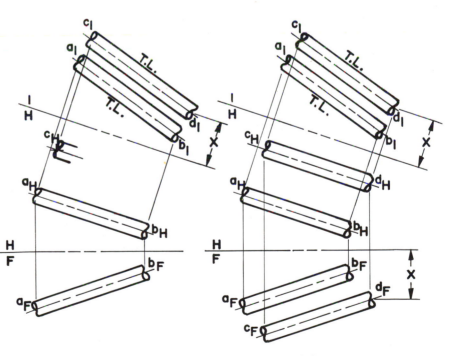

(a)	(b)	(c)

Figure 5-4. Construction of a line parallel to an existing line.

end of a 15-inch shaft, **CD**, that runs parallel to shaft **AB**. At (b), auxiliary view 1 has been drawn showing the true length of shaft **AB**. Since the shafts are parallel, shaft **CD** appears parallel to **AB** and in its true length (15 inches) in this view. Finally, at (c), the newly found point D is projected back to the top and front views. Because **AB** and **CD** are parallel, and appear parallel in all three views, d$_H$ can be found by simple projection, and d$_F$ can be found by projection and measurement. (Note: Distance **X** must be the same in view 1 and in the front view.) If you wanted to obtain the true distance between these shafts, what would you do next?

5.3 PERPENDICULAR LINES

Although two lines may make a variety of angles with each other, perpendicular lines occur often and deserve separate study. Lines are perpendicular, when there is a 90° angle between them. Lines may be perpendicular whether they are intersecting or nonintersecting. Figure 5-5 shows two pieces of duct that intersect at 90° to each other, while Figure 5-6 illustrates the same two types of duct that are nonintersecting, yet still perpendicular to each other.

Whether they are intersecting or nonintersecting, **lines that are perpendicular to each other in space will appear at right angles to each other in any view that shows either or both lines in true length**. The only exception is when one line appears as a point. Figure 5-7 shows a wedge with intersecting lines **AB**, **BC**, and **BD** labeled. That lines **BC** and **BD** are perpendicular to each other is seen in the top view where both lines appear in true length and the right angle is evident. Lines **BD** and **AB** are perpendicular as can be seen in the top and front views. In both views, line **BD** is drawn true length, and so the right angle is seen. Notice that line **AB** is not true length in either view. What other observations can you make about perpendicular lines on this object?

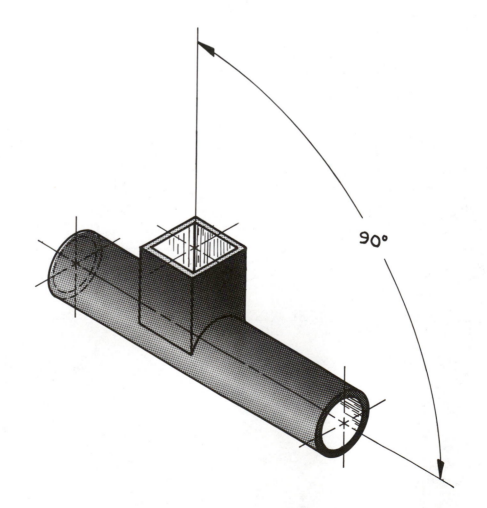

90°

Figure 5-5. Perpendicular lines: Intersecting ducts.

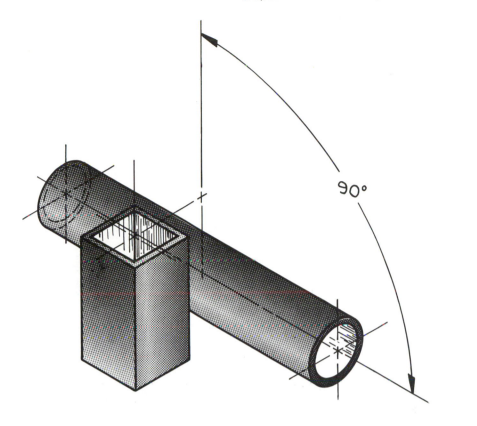

Figure 5-6. Perpendicular lines: Nonintersecting ducts.

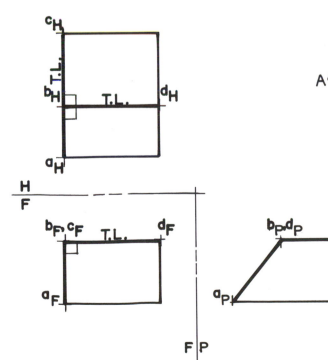

Figure 5-7. Perpendicular lines on a wedge.

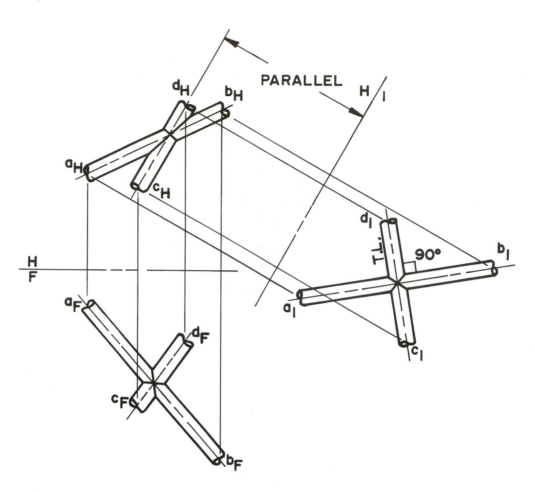

Figure 5-8. Establishing perpendicularity: Intersecting lines.

A more general situation illustrating two intersecting perpendicular pipes (**AB** and **CD**) is shown in Figure 5-8. The angular relationship between the two pipes cannot be seen in the horizontal or front views. This is because neither pipe appears true length in either view. In auxiliary view 1, which has been selected so that fold line H/1 is parallel to one of the pipes ($c_H d_H$), pipe **CD** appears true length ($c_1 d_1$) and the right angle between the pipes is seen.

The same principles and procedures apply with nonintersecting lines. Figure 5-9 shows two views of nonintersecting perpendicular shafts, **AB** and **CD**. To see their perpendicularity, fold line H/1 is drawn parallel to the horizontal projection of **CD**, causing **CD** to appear true length in view 1. **AB** and **CD** appear at right angles to each other in view 1, demonstrating that they are perpendicular in space. Only because the shafts are perpendicular to each other can view 2 be drawn showing shaft **AB** true length *and* showing an end view of shaft **CD**. Note that view 2 shows the true distance between the shafts.

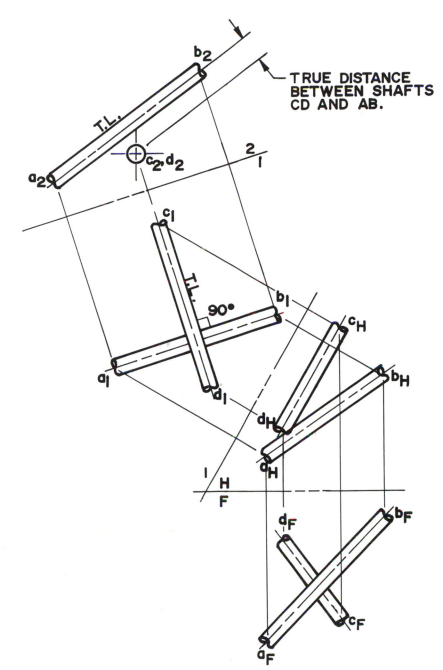

Figure 5-9. Establishing perpendicularity: Nonintersecting lines.

5.3.1 Shortest Distance (Perpendicular) from a Point to a Line—Line Method

The shortest distance from a particular point to a given line is a line drawn perpendicular to the given line. The shortest distance line will appear perpendicular when the given line is shown true length. Figure 5-10 shows the top and front views of a pipe centerline, **XY**, and a tank outlet at **Z**. A connecting pipe from the outlet at **Z** must join pipe **XY** with a standard 90° tee fitting at **W**. The required connecting pipe, **ZW**, is perpendicular to pipe, **XY**, and thus is the shortest line to it. View 1 is drawn adjacent to the horizontal view in order to show the pipe, **XY**, in true length. From z_1, the shortest line is drawn perpendicular to x_1y_1 to locate point w_1. Since **W** lies on **XY**, points w_H and w_F are located simply by alignment. To find the true length of the connecting pipe, **WZ**, view 2 is drawn parallel to w_1z_1 and is therefore perpendicular to x_1y_1.

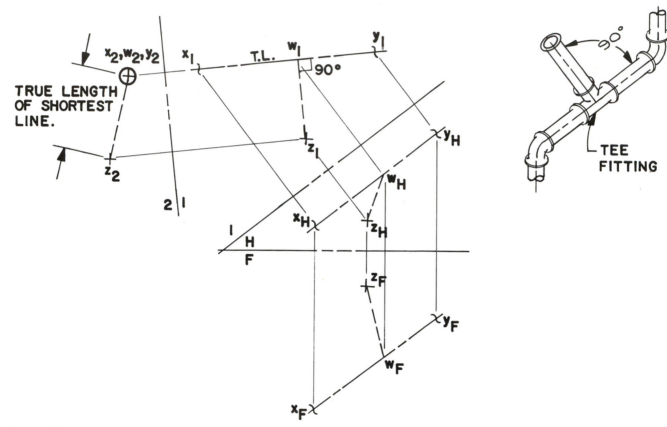

Figure 5-10. Construction of the shortest line from a point to a line: Line method.

5.3.2 Shortest Distance (Perpendicular) from a Point to a Line—Plane Method

As you learned in Chapter 4, a point and a line define a plane. If the plane is shown true size, then the shortest (perpendicular) line between the point and the line that defines the plane may be drawn in this view. Let's use a similar problem to the one solved in Figure 5-10, however, in Figure 5-11, it is solved employing the plane method. A plane, **XYZ**, is formed from pipe, **XY**, and point **Z**. The plane is found in an edge view in view 1. Then by viewing through a projection plane parallel to the edge view, the true shape of plane **XYZ** is shown in view 2. It is in view 2 that you can draw and measure the shortest line, a perpendicular, from point **Z** to the centerline of pipe, **XY**.

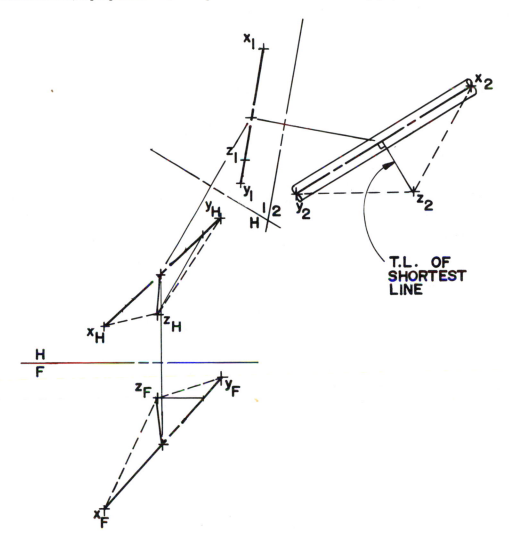

Figure 5-11. Construction of the shortest line from a point to a line: Plane method.

CHAPTER PROBLEMS

Notice that with each problem in this chapter, you are learning new applications of several of the fundamental views. You are moving into more varied and advanced aspects of descriptive geometry.

Problem 1: Given two views of pipelines **AB** and **CD**. Determine if they are parallel in space. Scale: $\frac{1}{8}" = 1'\text{-}0"$

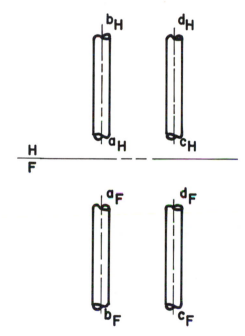

Problem 2: You are asked to draw a pipeline, **WZ**, parallel to the existing pipeline, **XY**. They are equal in length and are both 2 inches in diameter. What is the clearance (distance) between them? Scale: 1" = 10"

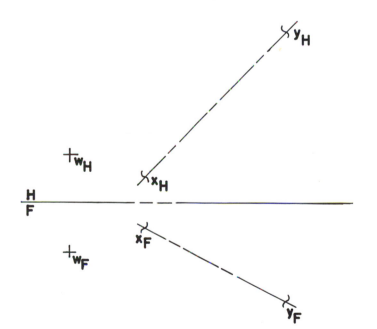

Problem 3: Line **VW** is one side of a parallelogram, and point **Y** is the midpoint of the opposite parallel side. Draw the top and front views of the parallelogram. Scale: ½

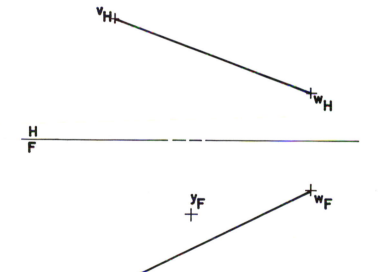

Problem 4: Line **CD** is the centerline of the axle of a 2-inch diameter wheel whose center is at the midpoint of **CD**. The wheel has six spokes, one pair being horizontal. Draw the top and front views of the wheel with spokes. You may disregard the wheel thickness. Scale: Full

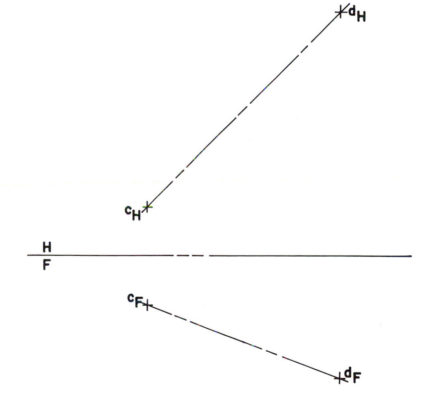

DR. BY:	COURSE & SEC:	SCALE:	DATE:

Problem 5: A hoist is supported by an A-frame, whose two legs are perpendicular to the load-bearing beam, **XY**, and to each other. Draw the center lines of the A-frame legs from point **Z** to the floor. Scale: 1" = 50"

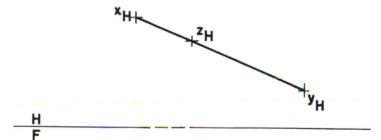

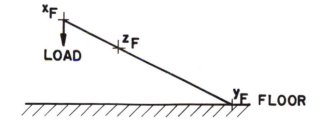

Problem 6: Determine the shortest airway to be dug from a point **X** on the earth's surface to the centerline of a tunnel, **YZ**.

Find the true length, grade, and bearing of the airway. Scale: 1" = 50'

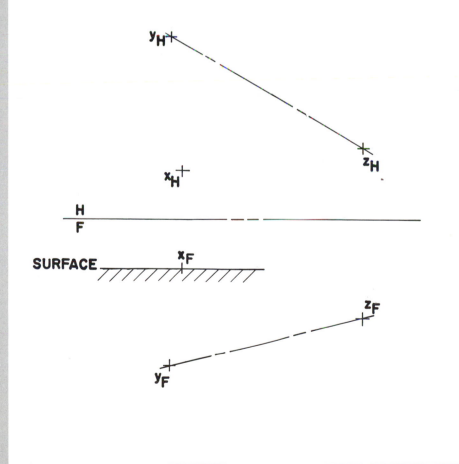

Problem 7: Line **AB** is the centerline of a gas pipe running through a mechanical room of a building. A Tee fitting is to be inserted along this pipe to allow for a straight connecting pipe to the boiler at **C**. Determine the true length of the shortest connecting pipe from **C** to **AB**. Show the position of the connector in all the views. Scale: 1" = 20"

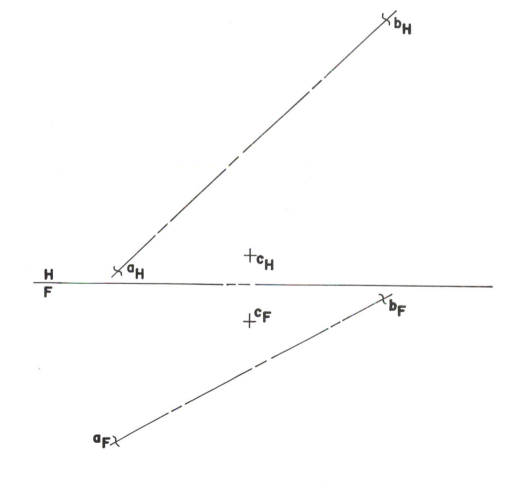

| DR. BY: | COURSE & SEC: | | SCALE: | DATE: |

Problem 8: You have a plot plan and front view of a corner lot in a residential area. **AB** and **BC** represent the centerlines of the existing water main. If the water meter at the house is located at **D**, what would be the length of the shortest pipe to the existing water line? Show the pipe in all views. Scale: 1" = 60'

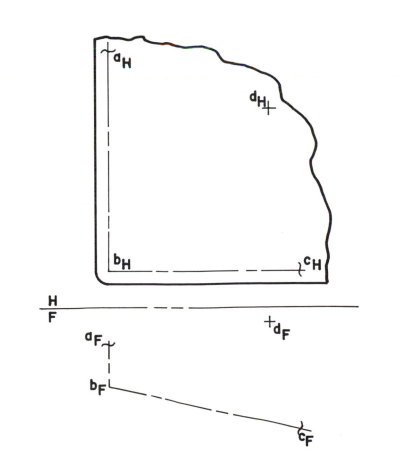

DR. BY:	COURSE & SEC:		SCALE :	DATE :

Problem 9: A length of 4-inch square tubing is used as a brace and is anchored to the floor at **A**, 81 inches in front of a wall. The tubing is attached to the wall 7 feet above the floor (directly behind the floor anchor). From a point 8 feet, 4 inches to the right of the corner and 11 inches above the floor, a length of 4-inch square tubing used as a stiffening brace, runs perpendicular to the main brace. The drawing shown is in a scale of ¼" = 1'-0". On a C-size drawing sheet, use a scale of ½" = 1'-0" to determine the true length of the main brace, and the true length of the stiffening brace. How far from the floor, measured along the main brace, is the connection? Dimension both the main brace and the stiffening brace for fabrication.

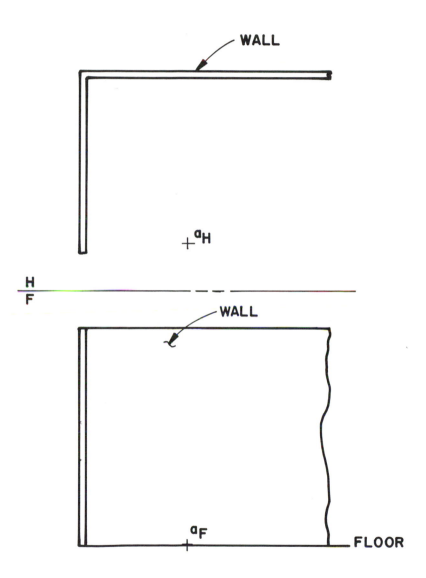

WALL

+ ^{a}H

$\dfrac{H}{F}$

WALL

^{a}F

FLOOR

Problem 10: Pinion gear **X** drives spur gear **Y**. The pitch diameters of the mating gears are tangent. **C** is a point on the pitch diameter of gear **X**. Lines **AB** and **EF** represent the center lines of the gear shafts. The scale of the drawing shown is ¼. In full scale, on a C-size drawing sheet, determine the pitch diameters of gears **X** and **Y**. Show the pitch diameters as phantom lines in all views.

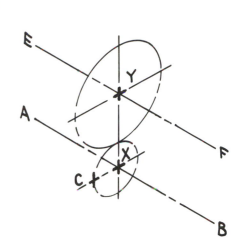

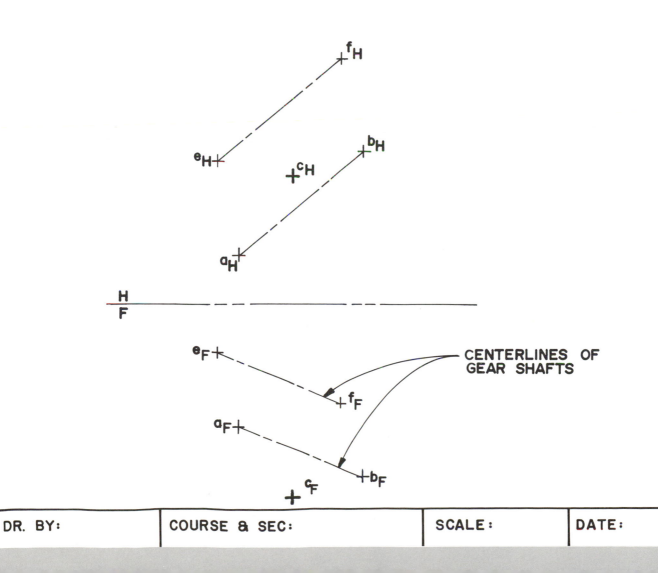

CENTERLINES OF
GEAR SHAFTS

DR. BY:	COURSE & SEC:	SCALE:	DATE:

CHAPTER FIVE

TEST

1. Define parallel. Show an example that illustrates your answer.

2. Define perpendicular. Show an example that illustrates your answer.

3. If lines are parallel in space, they will appear _____ in all views, except two. List them.

4. How must parallel lines appear in order to see the true distance between them?

5. Indicate whether the following statements are true or false.

  T F a. All horizontal lines are not parallel to each other.

  T F b. If lines appear parallel in any two adjacent views, they are parallel in reality.

  T F c. Perpendicular lines do not have to intersect.

  T F d. If two lines are perpendicular in actuality, they will appear parallel in any orthographic view.

  T F e. Frontal and horizontal lines are always perpendicular to each other.

6. If two lines are intersecting and perpendicular to each other, in what view (or views) will you see the right angle between them?

7. If two lines are perpendicular and nonintersecting, how will they appear when you can see the true distance between them?

8. When the shortest distance from a point to a given line is seen true length, how is the given line seen?

9. There are two methods of finding the shortest distance between a point and a line. List them.

10. Explain the difference in procedure between the two methods listed in your answer to question 9.

CHAPTER

SIX

Intersecting and Nonintersecting Lines

6.1 INTERSECTING LINES

When lines are intersecting, the point of intersection is a point which lies on both lines. Figure 6-1 shows two intersecting cables, **AB** and **CD**, that intersect at the cable junction, point **X**. You know the two cables intersect because both the horizontal and front views of point **X** are on the same projection line. Point x_H is directly above x_F, proving that point **X** is a point on both cables and that they intersect.

Figure 6-2 illustrates the special case of a profile alignment. Cables **AB** and **CD** appear to intersect, but without the additional view (right-profile) the intersection is not

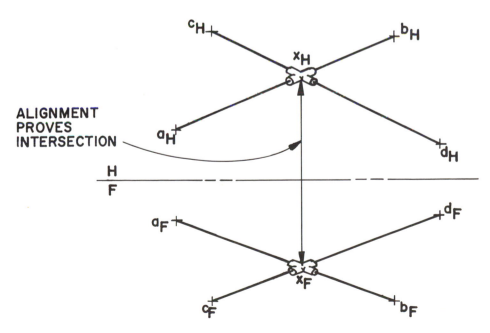

Figure 6-1. Intersecting cables.

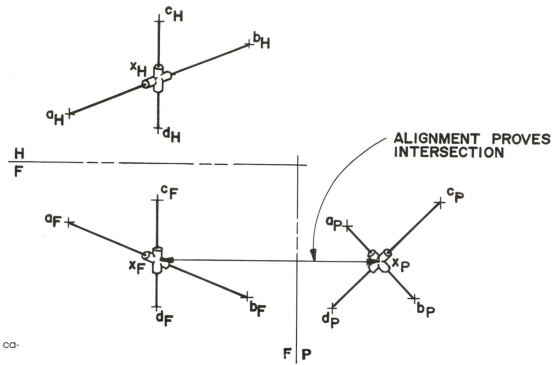

ALIGNMENT PROVES INTERSECTION

Figure 6-2. Intersecting cables: Profile lines.

certain. With the right-profile view, it can be seen that x_F and x_P align with each other, thus proving that cables **AB** and **CD** intersect at point **X**.

Conversely, when lines are nonintersecting, the apparent point of intersection does not align between adjacent views, which proves that they do not intersect. Point x_H in Figure 6-3 does not align with a point of intersection in the front view, and point y_F does not align with a point of intersection in the horizontal view. The pipe is below and in front of the belt.

6.2 VISIBILITY OF LINES

When two lines do not intersect, it is important to know which line is visible and which is hidden. When you are looking down on the top view, whichever line is highest in elevation will be visible. Looking toward the front view, whichever line is closer to the front projection plane will be visible.

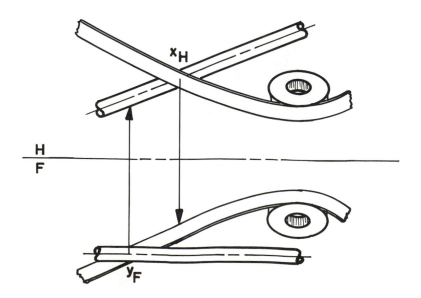

Figure 6-3. Nonintersecting lines.

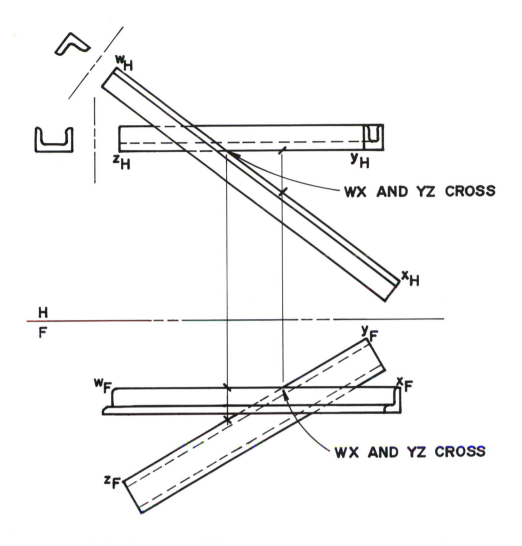

Figure 6-4. Visibility of lines.

In order to determine whether angle **WX** or channel **ZY** is visible in the top view of Figure 6-4, a projection line is dropped down from the place where w_Hx_H and y_Hz_H cross in the top view. Line w_Fx_F is closer to the top (higher) than y_Fz_F, therefore the angle **WX** is visible in the horizontal view. To determine the visibility in the front view, a projection line is drawn upward from the crossing of w_Fx_F and y_Fz_F. Line w_Hx_H is closer to the front (closer to the observer) than y_Fz_F, therefore the angle **WX** is visible in the front view and the channel, **YZ**, is hidden.

This method is a general one and is applied not simply to line problems, but to all situations where visibility is in question. At (a) in Figure 6-5, two views of a tetrahedron are shown, with the visibility of the diagonal lines undetermined. To determine the visibility in the top view, project to the front view the point where e_Hg_H crosses f_Hh_H to see which line is higher. Observe that e_Fg_F is higher than f_Fh_F, and therefore it is visible in the top view as shown at (b). Visibility in the front view is determined by projecting to the top view the point where e_Fg_F crosses f_Fh_F and then observing which line is closer to the front projection plane. Line f_Hh_H is closer to fold line H/F than e_Hg_H, therefore line f_Fh_F is visible in the front view as can be seen in the front view at (b).

6.3 LOCATION OF A PLANE THAT CONTAINS ONE LINE AND IS PARALLEL TO ANOTHER LINE

At times it is necessary to construct a plane that contains, or includes, a given line and which is also parallel to another given line. **By geometry, when a line is parallel to any line on a plane, it is parallel to that plane.** The use of this axiom will be seen in section 6.4. Figure 6-6 shows two nonintersecting, nonparallel lines, **AB** and **CD**, and a

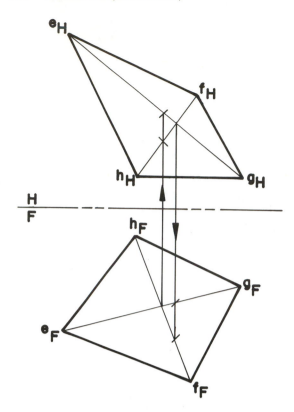

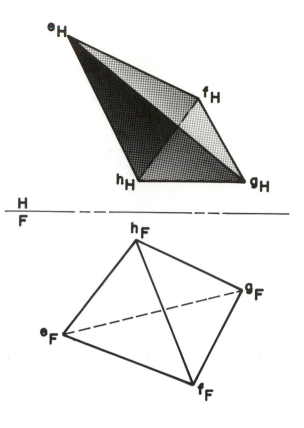

Figure 6-5. Visibility of lines. **(a)**

(b)

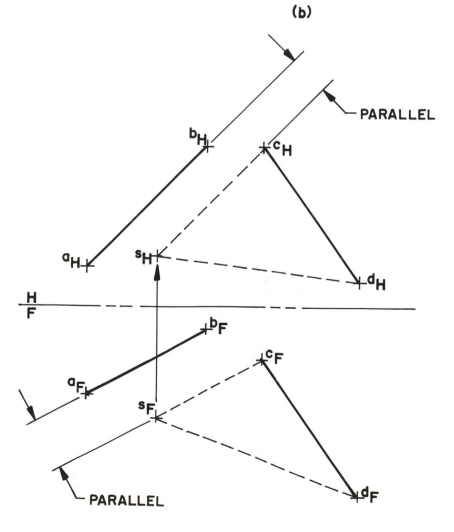

Figure 6-6. A plane containing one line and parallel to another.

plane, **CSD**, which contains line **CD**. Plane **CSD** is parallel to line **AB**, because **CS** is parallel to **AB**.

The construction of this plane begins with a line drawn from c_F parallel to $a_F b_F$. The line ends arbitrarily at s_F, and then s_F is connected to d_F. In the horizontal view, s_H is found by projecting s_F onto a line drawn parallel to $a_H b_H$ from c_H. Then s_H is connected to d_H to complete the horizontal view of the plane. Study this construction carefully, and try to visualize it in space.

6.4 SHORTEST (PERPENDICULAR) DISTANCE BETWEEN TWO NONINTERSECTING, NONPARALLEL LINES

You already know what the shortest distance from any point to a line is the perpendicular distance between them. Likewise, the shortest distance between two lines must be a perpendicular line between both lines. There is only one position in space where it is possible to have a line perpendicular to two other lines. There are two methods of solution to this type of problem, the line method and the plane method.

6.4.1 The Line Method

In Figure 6-7 two cables, **AB** and **CD**, on an aircraft are shown. It is necessary to connect them with the shortest possible stabilizing cable, **XY**. Cable **XY** is found by drawing view 1 showing the true length of cable **AB**. View 2 is constructed to show cable **AB** as a point. Cable **CD** is also shown in view 2, and the common perpendicular is drawn to cable **CD** from the point view of **AB**. Note that the true length of cable **XY** is shown in view 2. Stabilizing cable **XY** is projected back to view 1. Point **Y** is projected onto cable $c_1 d_1$, and point **X** is drawn at right angles to cable $a_1 b_1$. Remember, a perpendicular to a line will show perpendicularity in the view that shows the true length of the line. Both ends of the stabilizing cable, **XY**, may be projected to the horizontal and front views by simple alignment.

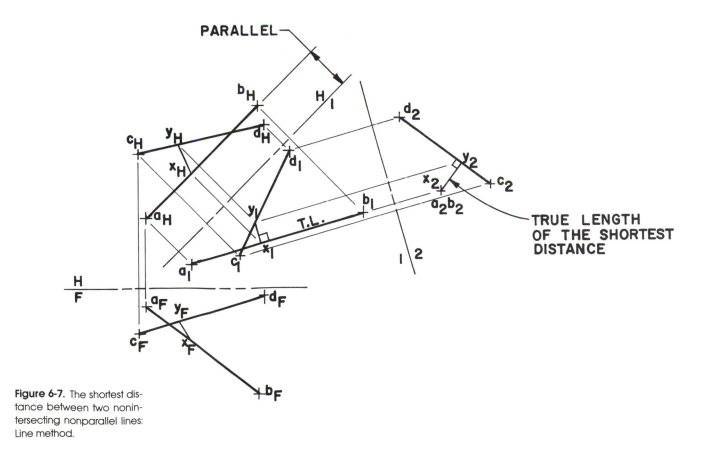

Figure 6-7. The shortest distance between two nonintersecting nonparallel lines: Line method.

6.4.2 The Plane Method

Let's solve the same problem, using the plane method (see Figure 6-8). First it is necessary to construct a plane which contains cable **CD** and is parallel to cable **AB**. To do this, $c_F e_F$ is drawn parallel to $a_F b_F$. Notice that $e_F d_F$ is drawn parallel to fold line H/F. Therefore, it will be true length in the horizontal view. The horizontal view of plane **CDE** is found by drawing a line from c_H parallel to $a_H b_H$ until it intersects a line projected from point e_F, which locates e_H. Then e_H is connected to d_H to complete the plane. Remember line $e_H d_H$ is true length. Fold line H/1 is drawn perpendicular to $e_H d_H$ causing plane **CDE** to be seen as an edge in view 1 and line $a_1 b_1$ to appear parallel to the edge view of **CDE**. Fold line 1/2 is drawn parallel to the edge view of plane $c_1 d_1 e_1$ and line $a_1 b_1$. View 2 shows the plane in its true shape and both cables in their true length. In view 2, where the two lines appear to intersect, the common perpendicular (the stabilizing cable) appears as point $x_2 y_2$. The true length of **XY** is shown in view 1. The stabilizing cable, **XY**, may be projected back to the horizontal and front views by simple alignment. When you compare Figures 6-7 and 6-8, stabilizing cable **XY** is the same length and in the same location in both top and front views.

Another frequent application of the shortest perpendicular distance technique is in the determination of clearance. The shortest distance could be the clearance needed for a valve in a piping system or the clearance required between power lines and a nearby metal stucture in an electrical power substation. The examples are many and varied.

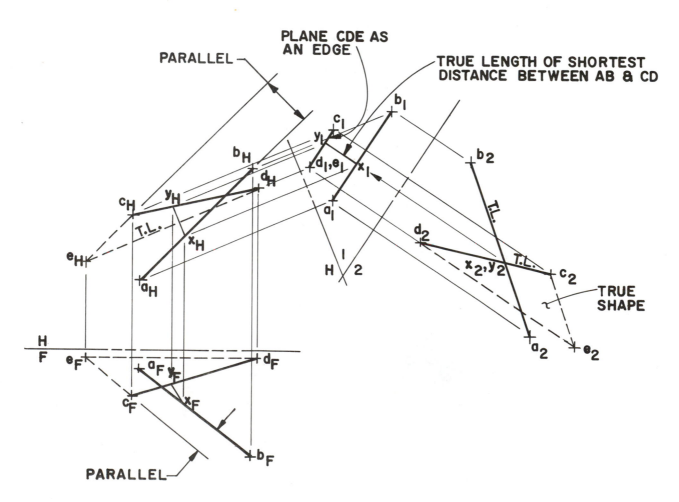

Figure 6-8. The shortest distance between two nonintersecting nonparallel lines: Plane method.

6.5 SHORTEST LEVEL (HORIZONTAL) LINE CONNECTING TWO NONINTERSECTING, NONPARALLEL LINES

A problem very similar to the one just described is one in which the required connecting line between two given lines is level, or horizontal, rather than perpendicular. Figure 6-9 shows the centerlines of two mine shafts, **KL** and **MN**. It is necessary to find the shortest level tunnel, **ST**, between them. First, a plane $m_F n_F p_F$ is created containing mine shaft **MN**, with line $n_F p_F$ parallel to $k_F l_F$ and line $m_F p_F$ parallel to fold line H/F. Plane $m_H n_H p_H$ is found by drawing a line from n_H parallel to $k_H l_H$ until it intersects a line projected from p_F. Next, draw line $m_H p_H$, which will be true length.

Fold line H/1 is drawn perpendicular to $m_H p_H$ causing plane $m_H n_H p_H$ to appear as an edge view, with mine shaft $k_1 l_1$ parallel to it. Next, view 2 is found by drawing fold line 1/2 *perpendicular to fold line H/1*. In view 2 the connecting tunnel, $s_2 t_2$, will appear as a point where the two mine shafts appear to intersect. The position of the level tunnel is now fixed and is located in the other views by projection. Note that $s_1 t_1$ is parallel to fold line H/1, as is $s_H t_H$, and that $s_F t_F$ is level.

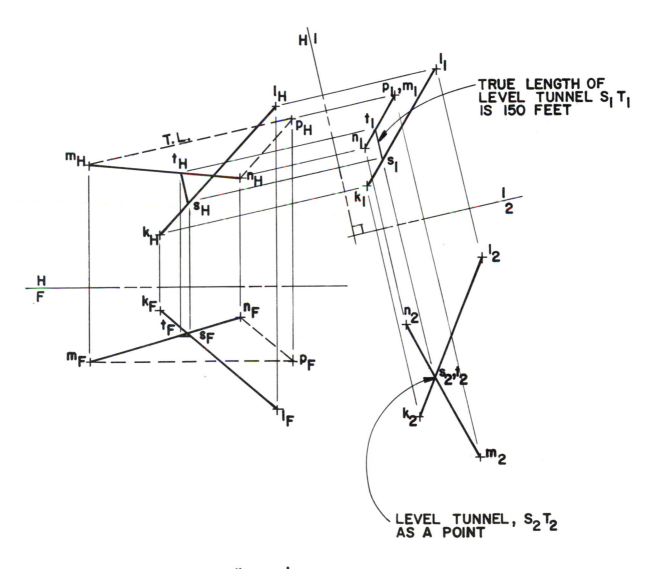

Figure 6-9. The shortest level (horizontal) line connecting two nonintersecting nonparallel lines.

Scale: 1" = 500'

6.6 SHORTEST LINE AT A GIVEN GRADE, OR SLOPE, CONNECTING NONINTERSECTING, NONPARALLEL LINES

A third type of problem similar to the others discussed in this chapter is one in which the required line is not level, or perpendicular, but instead must have a specified grade, or slope angle. In Figure 6-10 two pipes, **AB** and **CD**, are to be connected with a pipe at a −40 percent grade from **CD** to **AB**.

As in sections 6.4 and 6.5, a plane is drawn containing one of the lines and parallel to the other line. In auxiliary view 1, the pipes appear parallel. To position the shortest connecting pipe at a −40 percent grade from **CD** to **AB**, you must draw fold line 1/2 *perpendicular to a line drawn at a −40 percent grade from fold line H/1*. In view 2, the pipes appear to intersect and the connector appears as a point. When projected back to view 1, the connecting pipe is shown in its true length. The true length, centerline to centerline, measures 3 feet 4 inches.

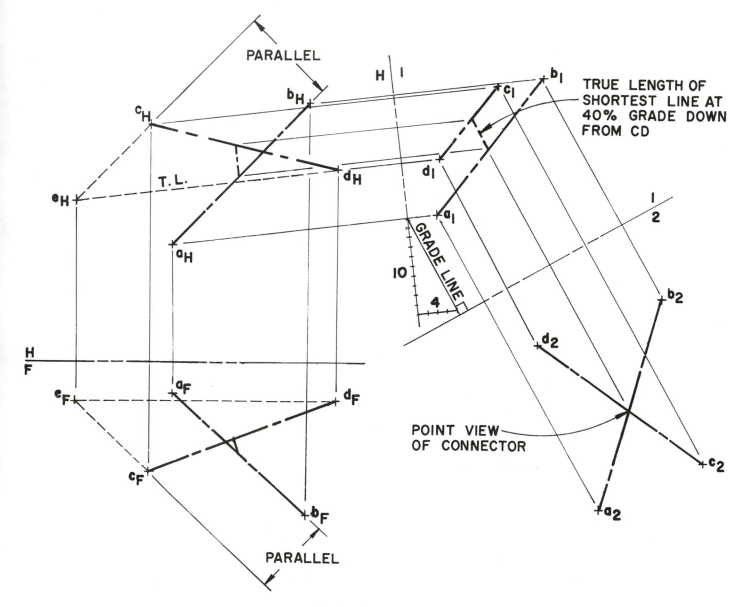

Scale: 1/8" = 1'-0"

Figure 6-10. The shortest line at a given grade connecting two nonintersecting nonparallel lines.

Let's try one more example of this type of problem. In this case (see Figure 6-11), pipes **WX** and **YZ** are to be connected with a standard 45° Y fitting, upwards from **YZ**. The procedure for solving this problem is as follows:

1. In the front view, draw a plane containing pipe **WX**, with a line parallel to pipe **YZ**, and with line, **VX**, parallel to fold line H/F. Project plane **VWX** to the top view.
2. Find an edge view of plane **VWX** in view 1.
3. Draw fold line 1/2 perpendicular to a line drawn at 45° upwards from fold line H/1.
4. In view 2, the pipes appear to intersect. The shortest connector at a 45° slope angle appears as a point in view 2, and in true length in view 1.

Examine Figures 6-10 and 6-11 carefully. Try to visualize the difference in orientation due to a 45° uphill slope versus a 40 percent downward grade.

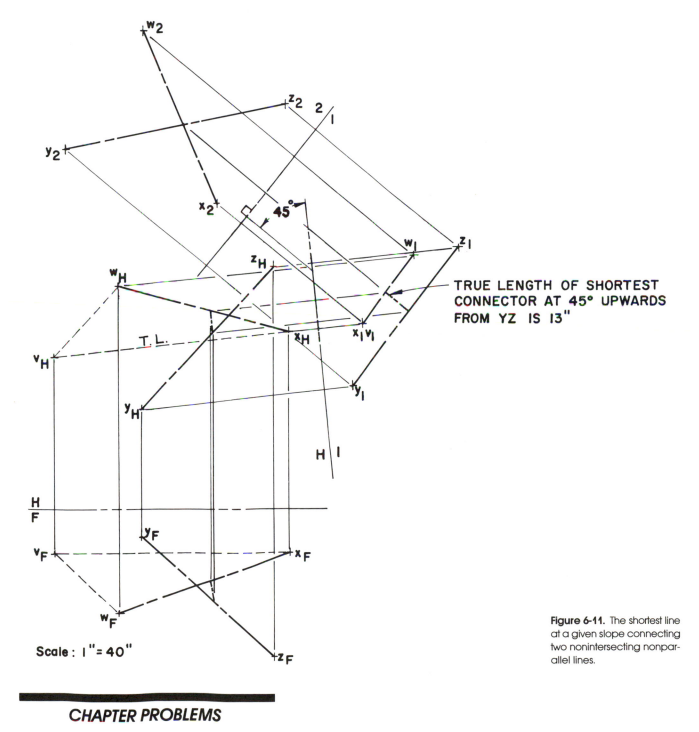

TRUE LENGTH OF SHORTEST CONNECTOR AT 45° UPWARDS FROM YZ IS 13"

Scale: 1" = 40"

Figure 6-11. The shortest line at a given slope connecting two nonintersecting nonparallel lines.

CHAPTER PROBLEMS

Problem 1: Show the correct visibility for the shafts below.

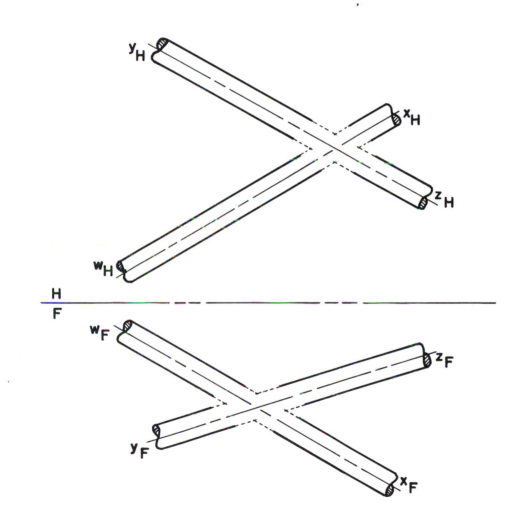

Problem 2: Complete the visibility of the structural angle and channel shown below.

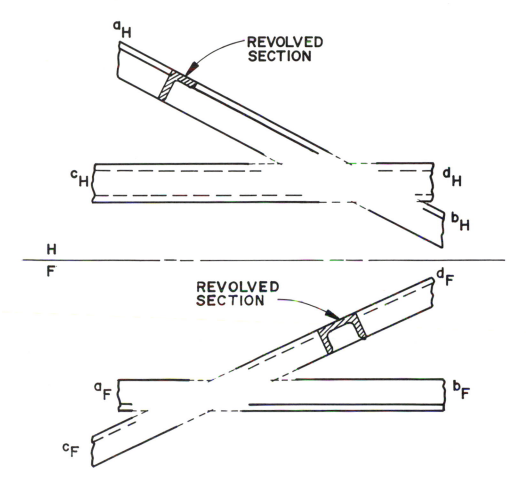

Problem 3: Three .25-inch diameter rods are shown. Construct
a profile view. Show the correct visibility in all views.

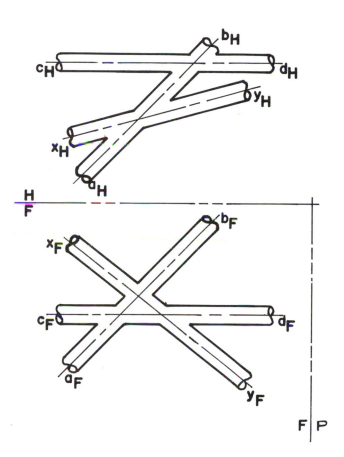

Problem 4: The design of a high-voltage transmission tower requires that the clearance between the high-voltage power lines and the structural members of the tower be known. Find the clearance between power line **AB** and structural member **CD**. Show this line of clearance in all views. Scale: 1" = 20'

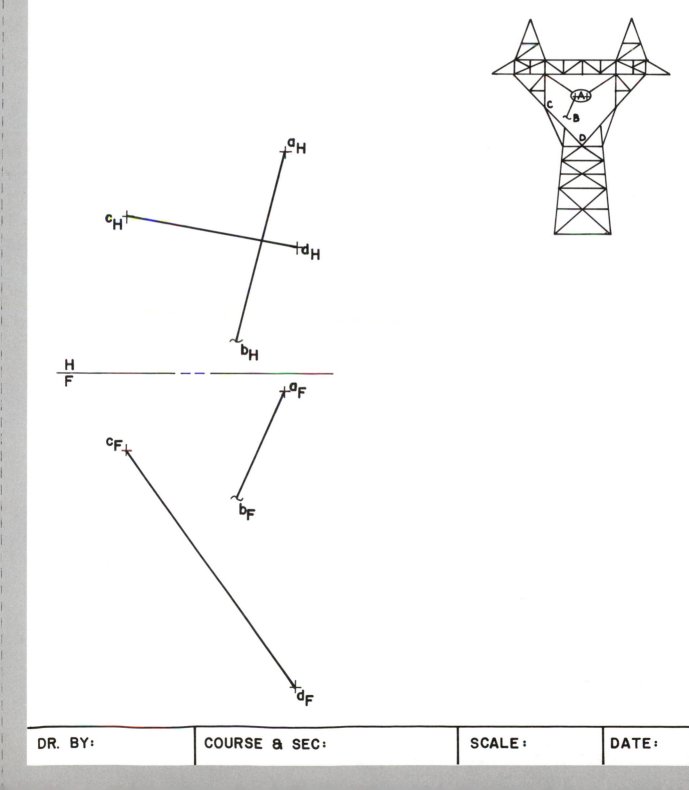

Problem 5: A vertical mast, **VW**, 18 feet high and a wire, **AB**, are located as shown. A guy wire from point **X** is to be fastened as high as possible to the mast at point **Y**, yet it must clear wire **AB** by 2 feet. Find the highest point on the vertical mast at which the guy wire can be fastened. Show the guy wire, **XY**, in all views. Scale: ⅛" = 1'-0"

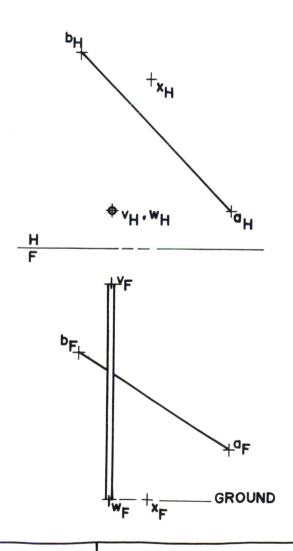

| DR. BY: | COURSE & SEC: | SCALE : | DATE : |

Problem 6: The clearance between two high voltage power lines must be at least 5 feet. If the present clearance is not sufficient, point **B** of line **AB** is to be lowered vertically until the necessary clearance is achieved. Show the new line **AB** in all views. (Note: The actual length of **AB** will change.) Scale: ¼" = 1'-0"

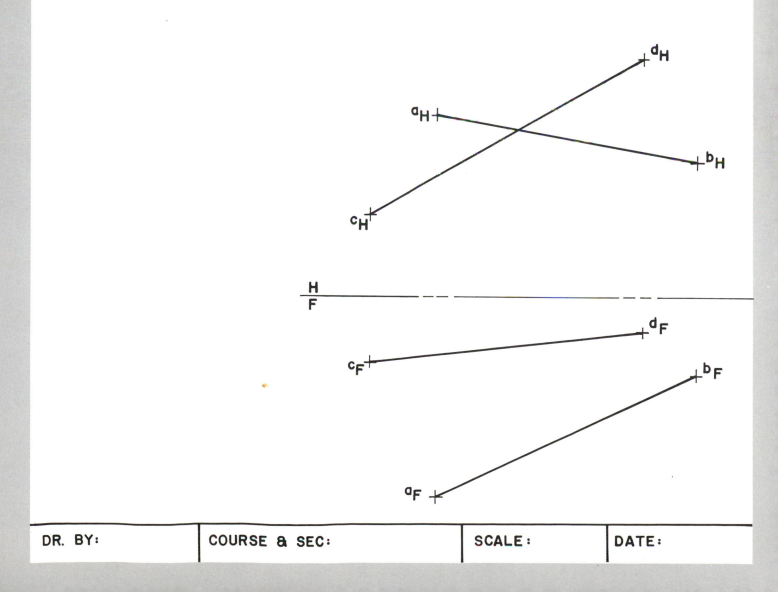

Problem 7: Lines **AB** and **CD** represent two control cables in an airplane. Find the minimum clearance between them. If the minimum clearance must be 4 inches, how far would you have to lower point **B** vertically to provide this minimum clear- ance between the cables? Point **B** is the only point that may be moved. Trace the problem, and complete your solution on a B-size drawing sheet. Scale 1" = 10'

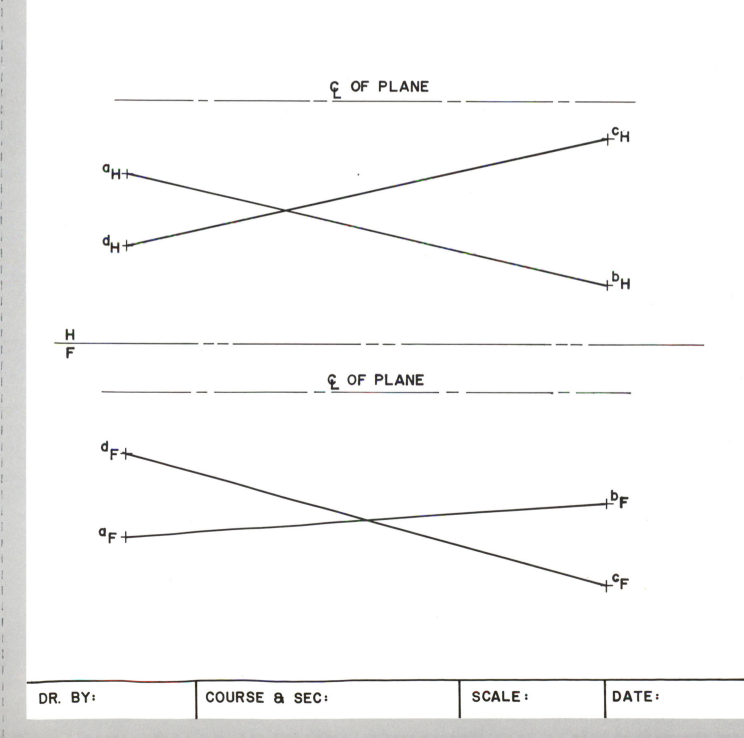

| DR. BY: | COURSE & SEC: | SCALE: | DATE: |

Problem 8: Line **AB** represents a new mine shaft, and **CD** represents an old one. It is desired to connect these two mine shafts with the shortest possible level tunnel. Find the true length and bearing of this level tunnel. Show the tunnel in all views. Scale 1" = 400'

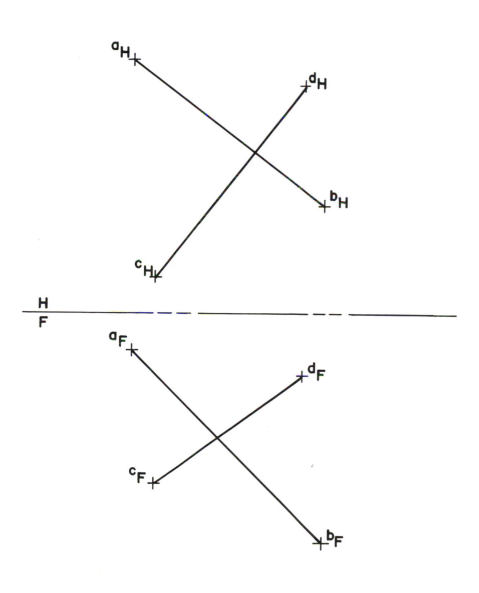

Problem 9: The centerlines of two pipes are shown in two views below. Find the true length of the shortest level connecting pipe (centerline) between them in order that a bypass valve may be installed. Scale the drawing shown below and show your solution on a B-size (11" × 17") drawing sheet. Be sure to show the centerline of the level pipe in all views. Scale: 1" = 30'

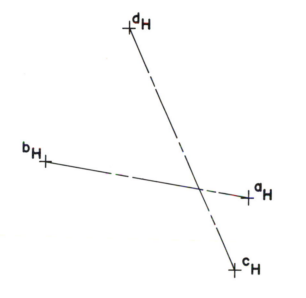

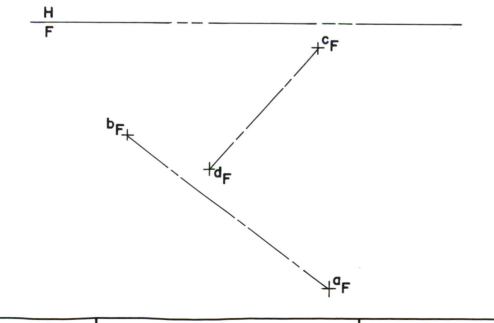

| DR. BY: | COURSE & SEC: | | SCALE: | DATE: |

Problem 10: Lines **AB** and **CD** represent the centerlines of two mining tunnels. Find the true length and bearing of the shortest level, connecting tunnel, **EF** between **AB** and **CD**. In addition, show a vertical escape shaft, **GH**, which connects **AB** and **CD**. Show the new construction in all views. Scale: 1" = 200'

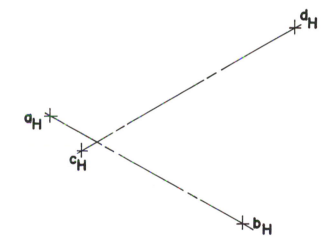

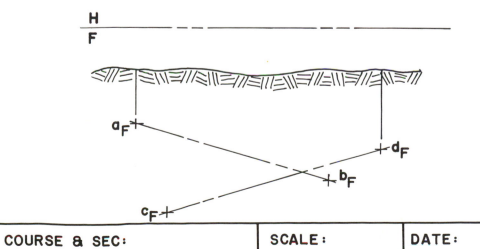

| DR. BY: | COURSE & SEC: | SCALE: | DATE: |

Problem 11: Lines **WX** and **YZ** are the centerlines of two natural gas lines. Find the shortest connecting pipe having a 20° downhill slope from **WX** to **YZ**. Show the connector in all views.

Trace the problem and complete your solution on B-size drawing sheet. Scale: 1" = 5'

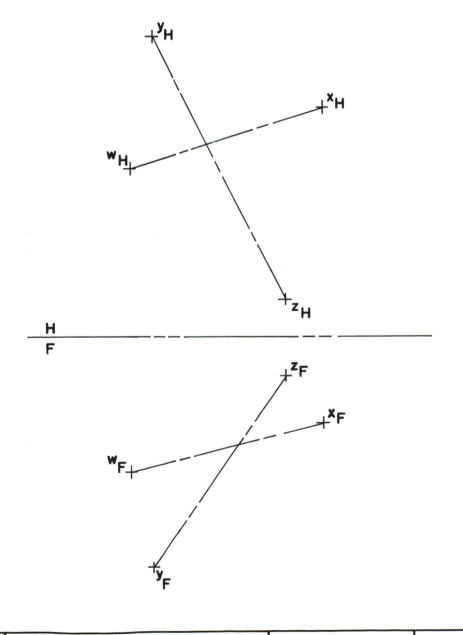

DR. BY:	COURSE & SEC:	SCALE :	DATE :

Problem 12: Two pipes are designated by their centerlines, **AB** and **CD**. It is necessary to connect them with the shortest pipe at a 29 percent grade down from **AB**. Find the true length of this connector, and show it in all views. Trace the problem and complete your solution on B-size drawing sheet. Scale: 1" = 20'

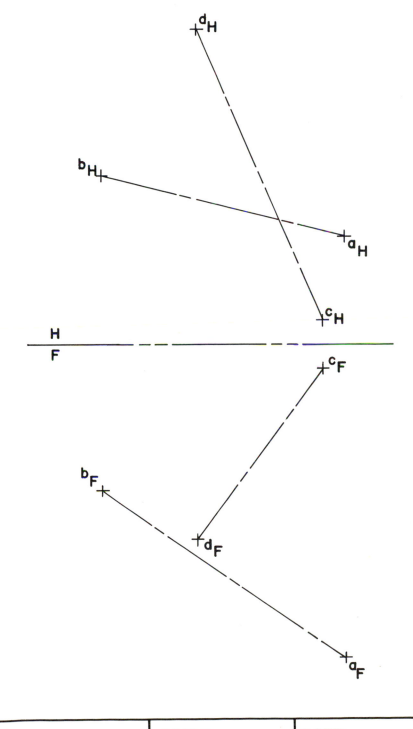

| DR. BY: | COURSE & SEC: | SCALE: | DATE: |

Problem 13: Two mining tunnels, **AB** and **CD**, are represented below. Point **Z** represents the opening of a new tunnel. Find the following: (a) The true length, bearing, and location of the shortest tunnel at a 35 percent grade from the opening at **Z** to the two tunnels, **AB** and **CD**. (b) The true length of the shortest vertical escape shaft from tunnels **AB** and **CD** to the surface. Scale: 1" = 300'

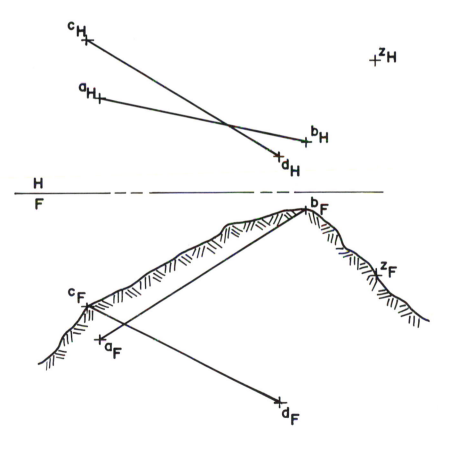

| DR. BY: | COURSE & SEC: | SCALE: | DATE: |

CHAPTER SIX

TEST

1. How can you test to determine if lines are actually intersecting?

2. Define the term *visibility* as it pertains to descriptive geometry.

3. Explain in your own words the process for determining the visibility of nonintersecting lines. (You may use a sketch to illustrate your explanation.)

4. Indicate whether the following statements are true or false.

   T   F   a. The outside lines of every view will be visible.

   T   F   b. Crossing edges that are approximately the same distance from the observer are tested for visibility at the crossing point.

   T   F   c. Visibility of inside lines in any view is always determined by reference to the top view.

   T   F   d. The corner of the object nearest to the observer will be hidden.

   T   F   e. To determine if a line is visible in the top view, simply observe the horizontal projection of the line.

5. Explain how to locate a plane which contains one line and is parallel to another line.

6. Given the horizontal and front positions of two pipes, explain how you would find the clearance between them using the line method.

7. How do you find the shortest horizontal connector between two oblique lines?

8. In which view(s) will you see the true length of the shortest level connector between two oblique lines?

9. Study the fold lines shown below. Is view 2 an inclined auxiliary view, or an auxiliary elevation view?

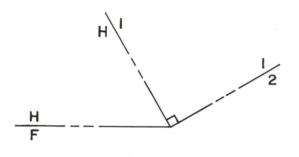

10a. Explain how you would solve the following problem: Lines **MN** and **RS** below represent two existing conveyor lines. They are to be connected by the shortest possible switching conveyor installed at a 10 percent grade downhill from **RS**.

10b. Illustrate your answer to question 10a.

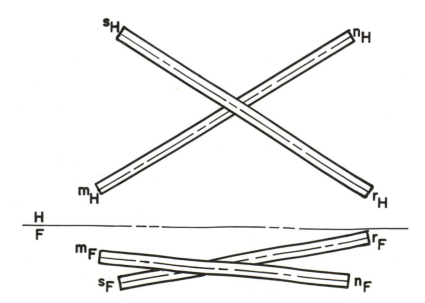

CHAPTER SEVEN

Plane Relationships

7.1 SHORTEST (PERPENDICULAR) LINE FROM A POINT TO A PLANE

The shortest (perpendicular) distance from a point to a plane will be seen in a view which shows the edge view of the plane. In such a view, the perpendicular line from the point to the plane can be drawn at right angles to the edge view of the plane. This perpendicular line appears true length in this view. In Figure 7-1, a foundry hopper is fed with fresh sand through a pipe conveyor from point **P** to the rear panel of the hopper. It is required to find the shortest pipe conveyor that can be used. To begin the solution, find the edge view of the rear panel (**ABCD**). This edge view is found in the right-profile view, because $a_H d_H$ and $c_H b_H$ are true length lines. Fold line H/P is drawn perpendicular to these true length lines. View P shows the edge view of the rear panel and a projection of point **P** (p_P). A perpendicular is drawn from p_P to the rear panel. This presents the true length of the centerline of the shortest pipe conveyor that can be used. This information determines the horizontal view. If $p_P r_P$ is true length, it *must* appear parallel to the H/P fold line in the horizontal view. Notice that $p_H r_H$ also appears perpendicular to the true length line, $a_H d_H$, in the horizontal view. This solution is just further application of the principles of perpendicular lines. How would you find the location and true size of the hole in the rear panel for the pipe conveyor?

7.2 SHORTEST GRADE LINE FROM A POINT TO A PLANE

The shortest distance from a point to a plane is a perpendicular to the plane, but at times the slope angle of the perpendicular is too steep for practical purposes. In an ore mine, for example, the entrance tunnel leading to the vein of ore (plane) should be short for economic reasons, but it should also have the best slope for the transportation of ore. Such a passage can be designed as the shortest grade (or slope) line from the point of entrance to the plane of the stratum of ore. For more mining terminology, see Chapter 12. In Figure 7-2, plane **ABCD** represents the upper plane (stratum) of an ore vein, and point **X** is a point representing the tunnel entrance. It is required to locate the shortest tunnel at a 20 percent grade. **To find the shortest grade line to the stratum, the**

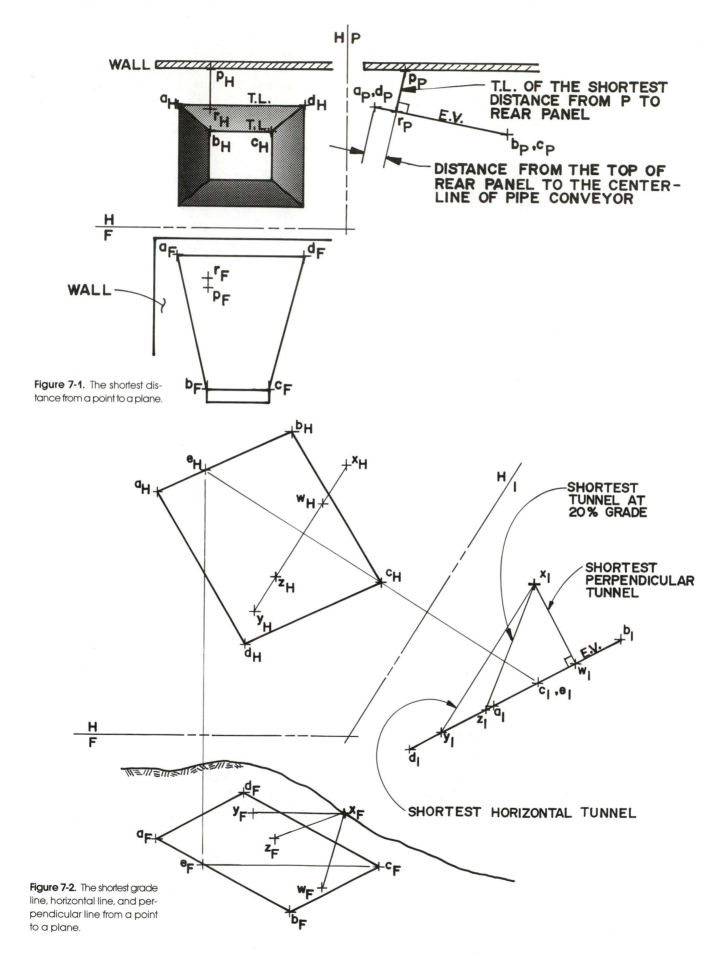

Figure 7-1. The shortest distance from a point to a plane.

Figure 7-2. The shortest grade line, horizontal line, and perpendicular line from a point to a plane.

stratum must appear as an edge in an elevation view (**view 1**). In view 1, it is possible to show the shortest horizontal tunnel, **XY**, the shortest perpendicular tunnel, **XY**, as well as the shortest tunnel at a 20 percent grade, **YW**, since each tunnel appears true length in view 1. Notice that all shortest lines, regardless of their grade, have the same bearing and are perpendicular to the true length line $c_H e_H$. Also remember that $x_H w_H$, $x_H y_H$, and $x_H z_H$ are all parallel to fold line H/1. Try to visualize this circumstance, and refresh your memory on why these observations are true.

7.3 THE ANGLE BETWEEN A LINE AND A PLANE

You have already learned that the slope angle of a line is the angle between the line and the horizontal plane. Remember that slope angle is seen in a view that shows the true length of the line, and that is an elevation view (one that shows the horizontal plane as an edge). In the same manner, **the angle formed by any oblique line and any oblique plane is seen in a view that shows the line true length in the same view that shows the plane as an edge.** Such a view is obtained from the following order: first, show an edge view of the plane, second, show the plane in true shape, and finally, show the plane again as an edge view where the line appears true length. In Figure 7-3 this procedure is applied to another foundry problem similar to that in Figure 7-1.

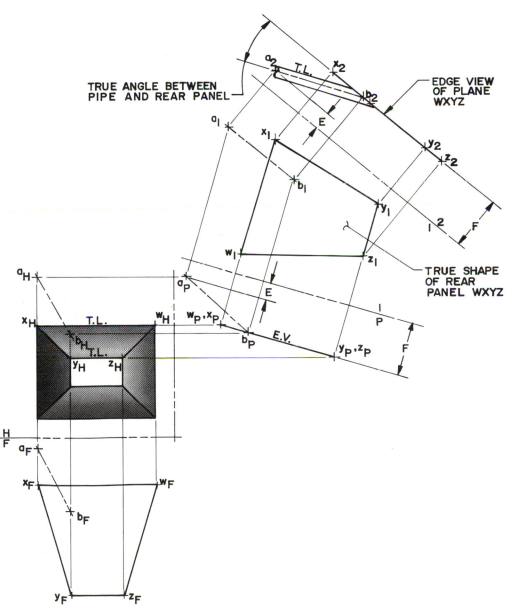

Figure 7-3. The angle between a line and a plane.

In this case, it is necessary to find the angle between the feeder pipe, **AB**, and the rear panel, **WXYZ**, of the sand hopper. In the right-profile view, an edge view of plane **WXYZ** is shown. Pipe **AB** is shown, although it is not true length. In view 1, the rear panel, **WXYZ**, is shown in true shape, and pipe **AB** is simply projected to this view. Finally, in view 2, the true length of pipe **AB** is seen along with the edge view of **WXYZ**. This condition has been accomplished by drawing fold line 1/2 parallel to a_1b_1, causing the rear panel to appear as an edge view again, and showing a_2b_2 in its true length in the same view. It is here that you can measure the true angle between the pipe and the rear panel of the hopper.

7.4 THE TRUE ANGLE BETWEEN TWO PLANES (DIHEDRAL ANGLE)

The angle that is formed by two intersecting planes is called the dihedral angle. **The dihedral angle is seen in its true size when both planes appear as edge views**. Both planes appear as edge views when the line of intersection between the two planes is shown as a point (see Figure 7-4). Figure 7-5 illustrates a hopper that feeds coal from an overhead storage compartment into a boiler stoker. The corners are to be reinforced with bent angles. It is necessary to find the true angle between planes **AEHD** and **ABFE** in order to design the bent angles properly. Carefully examine the procedure followed. First, view 1 is drawn in which the line of intersection, **AE**, between the two planes is found true length. Then fold line 1/2 is drawn perpendicular to true length line a_1e_1. In view 2, the line of intersection is seen as a point (a_2e_2) and the two planes are viewed as edges. The true size of the dihedral angle may be measured in view 2. As with all other examples in this text, try to visualize this problem by moving yourself through the four views.

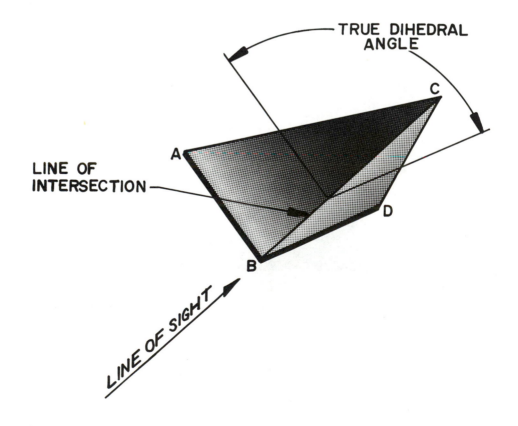

Figure 7-4. The true dihedral angle between two planes.

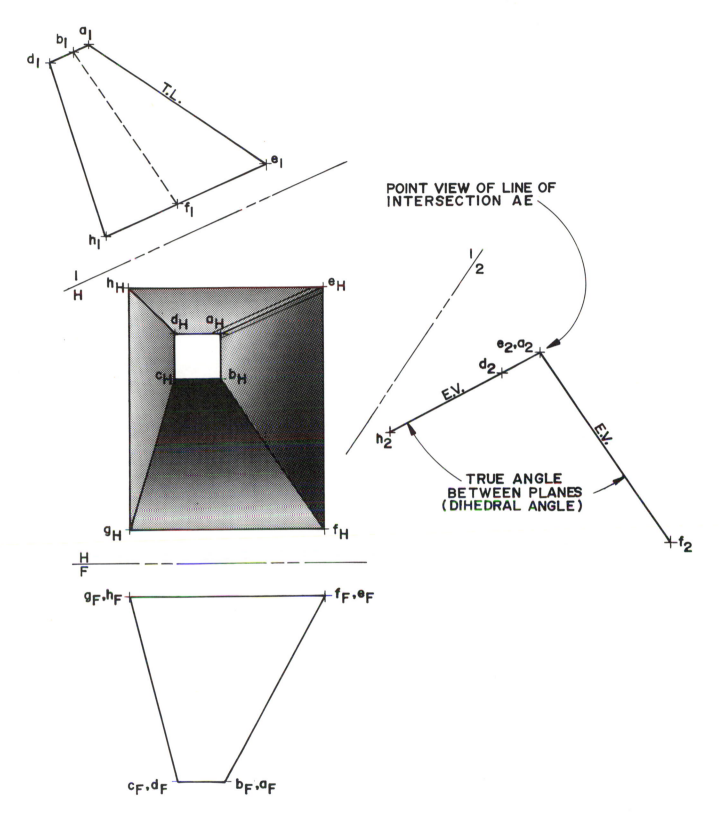

Figure 7-5. The true dihedral angle between two planes.

CHAPTER PROBLEMS

Problem 1: The roof of a house and a nearby pole are shown. Find the shortest possible guy wire to be anchored to roof plane **A** from the top of the pole at **X**. Dimension the true length of the guy wire. Show the guy wire in all views. Scale: 1" = 10'

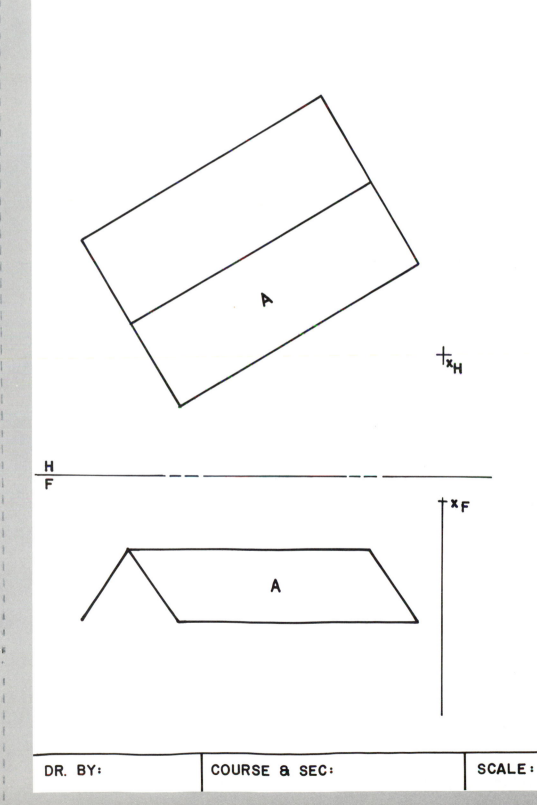

$\dfrac{H}{F}$

DR. BY:	COURSE & SEC:	SCALE :	DATE :

Problem 2: A storage hopper is located as shown below. Support braces made of 4-inch diameter pipe are to be fastened from points **X** and **Y** on the building walls to the nearest planes of the hopper. Find the true lengths of the two shortest support braces. (The braces will be welded in each position.) Select your own scale. Draw the given views, and your solution on a C-size (17″ × 22″) drawing sheet. Dimension the pipes for manufacture.

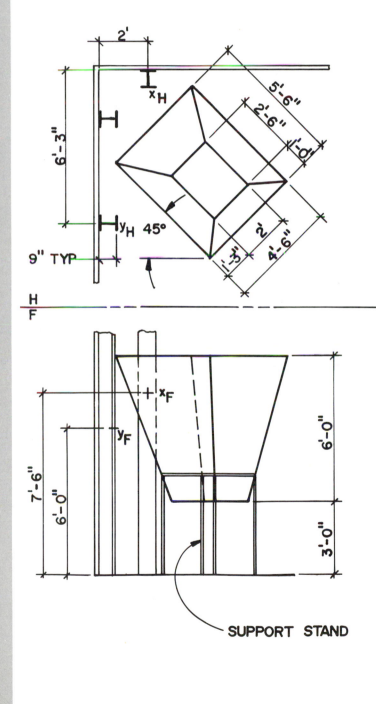

SUPPORT STAND

Problem 3: Two masts, **A** and **B**, stand next to a building. One guy wire from the top of each mast must be attached to the nearest roof surface and should be as short as possible. Show the two guy wires from **A** and **B** to the roof planes, **X** and **Y**, in all views and record the length of each. Scale: $1'' = 10'$

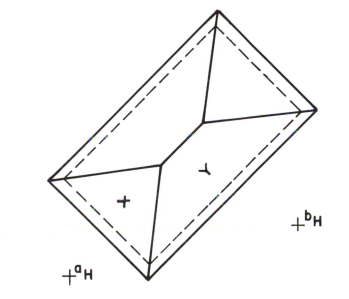

$\dfrac{H}{F}$

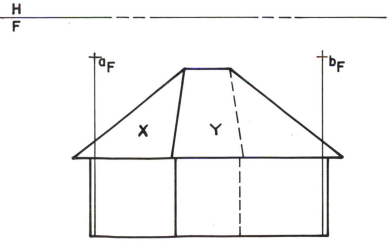

Problem 4: Points **X, Y,** and **Z** have been located on the upper surface (bedding plane) of a vein of ore. A new tunnel is to be dug from point **A** to the surface. Find the true length of the shortest tunnel at a −20 percent grade from **A** to the upper bedding plane. Show the tunnel in all views. Scale: 1" = 400'

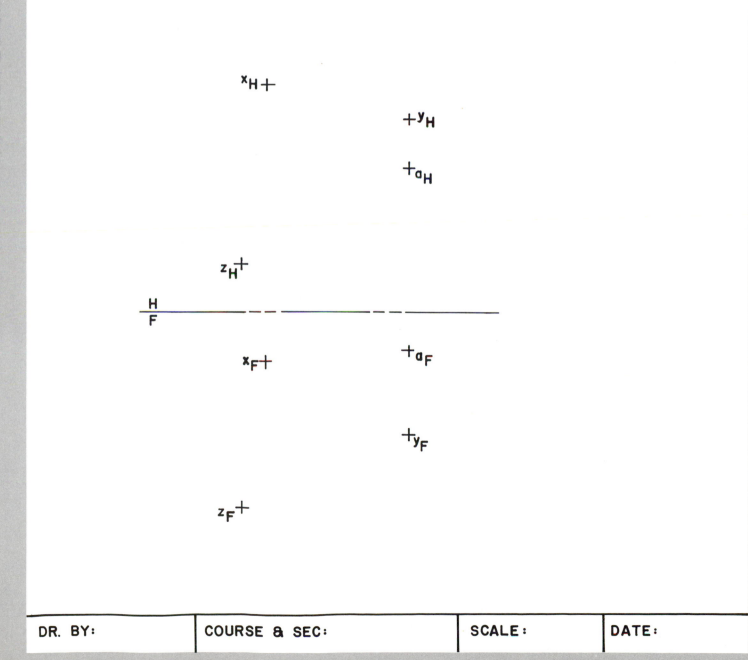

^{x}H+

+^{y}H

+^{a}H

^{z}H+

H
─
F

^{x}F+

+^{a}F

+^{y}F

^{z}F+

| DR. BY: | COURSE & SEC: | SCALE: | DATE: |

Problem 5: A metal tank is to fit below decks in the forward compartment of a ship. The tank has a level top, vertical sides, and a sloping bottom, **WXYZ**. From point **A** at the nearby bulkhead, a pipe must connect to plane **WXYZ**, and the pipe must have a 15° slope and a bearing of N 45° W from **A**. Find the true length of the centerline of the pipe. Show the pipe in all views. Scale: ⅜" = 1'-0"

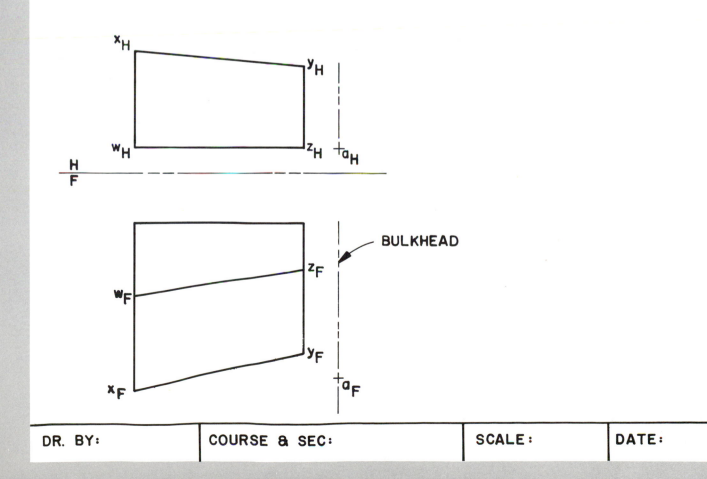

| DR. BY: | COURSE & SEC: | SCALE: | DATE: |

Problem 6: Advertising is to be placed on a signboard, **WXYZ**. The metal support frame is shown in dashed lines. From points **A** and **B**, two horizontal braces to the signboard are to be installed. Find the true angle these horizontal supports make with the signboard. Show the supports in all views. Scale: 1" = 20'

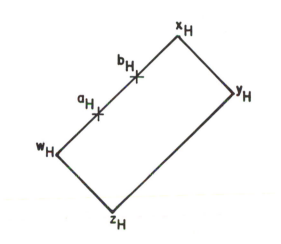

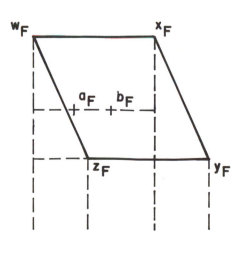

| DR. BY: | COURSE & SEC: | SCALE: | DATE: |

Problem 7: Line **MN** is the center of a pipe that intersects the side of a hopper at **M** and the floor at **N**. Find the angle the pipe makes with the hopper side and with the floor. Trace this problem and complete your solution on a B-size drawing sheet. Scale: $\frac{1}{2}'' = 1'\text{-}0''$

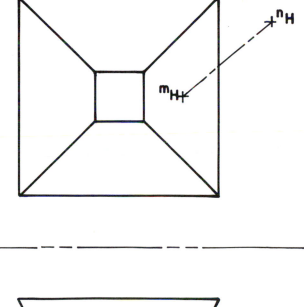

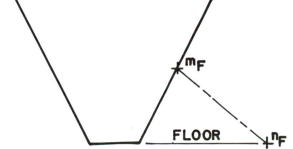

Problem 8: A mirrored surface, **ABCD**, is situated as shown below. Light ray **XY** will strike it as indicated. Find the angle of reflection of the ray with the mirror. Show the reflected ray, **YZ**, and all views. Scale: 1" = 20"

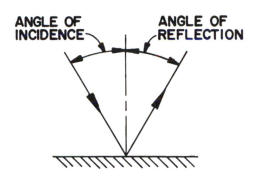

ANGLE OF INCIDENCE

ANGLE OF REFLECTION

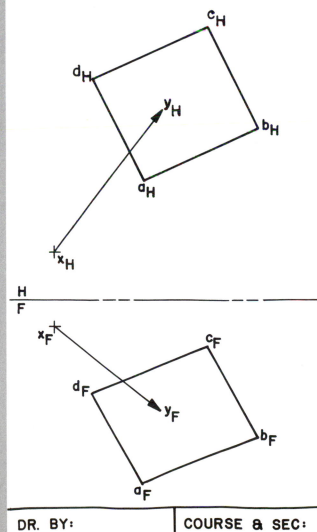

Problem 9: A ham radio antenna is supported by two support rods, **AB** and **AC**, attached to the roof of a house.

Find the angles made by the support rods and the roof.
Scale: ⅛" = 1'-0"

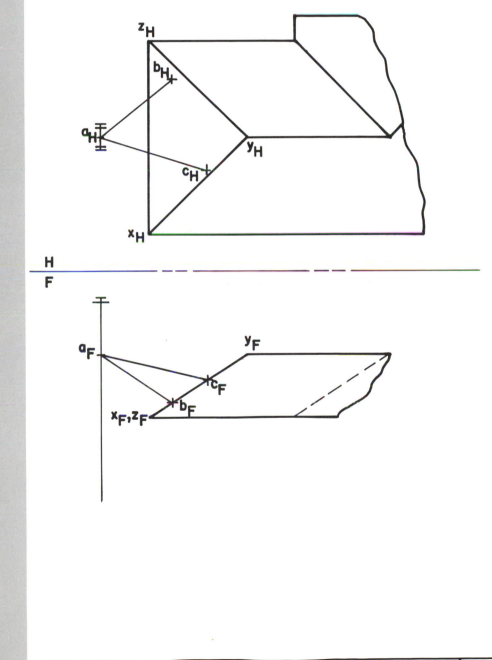

H
F

Problem 10: The sloping sides of the concrete bridge footing each have the same slope. Corners such as those between surfaces **X** and **Y** and surfaces **Y** and **Z** are usually covered with steel angle iron embedded in the concrete. Determine the angle to which each of these special iron plates must be bent. Scale: $\frac{1}{8}'' = 1'\text{-}0''$

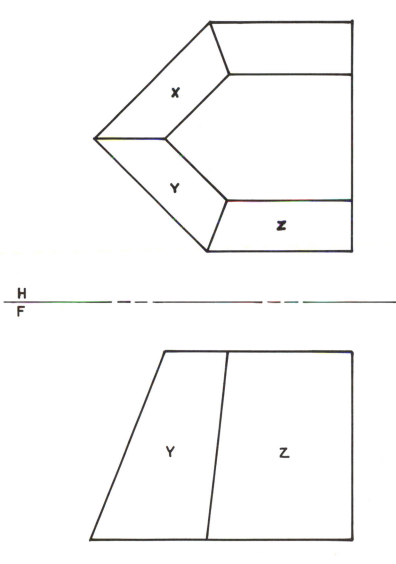

H / F

| DR. BY: | COURSE & SEC: | SCALE: | DATE: |

Problem 11: The structural members of one leg of an open tower are shown below. Special fittings are being designed for the joints. Find the angle between the planes **ABC** and **ABG**, and between planes **ABC** and **DEF**. (Note: **DE** is in the plane **ABC**.) Complete your solution on a C-size (17" × 22") drawing sheet using a scale of ¼" = 1'-0".

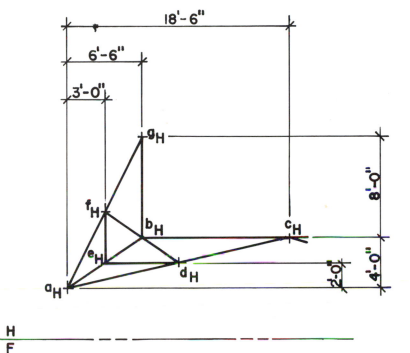

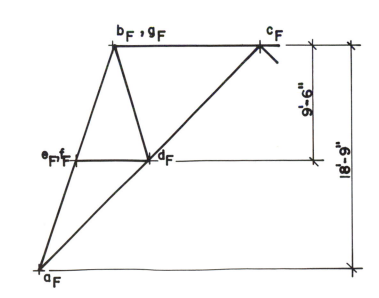

| DR. BY: | COURSE & SEC: | SCALE: | DATE: |

Problem 12: A sheet metal chute is to be installed between the second and third floors of a building. Find the true size of the angle between planes **A** and **B**, **B** and **C**, **C** and **D**, and **D** and **A**. Use a C-size (17" × 22") drawing sheet and a scale of ½" = 1'-0".

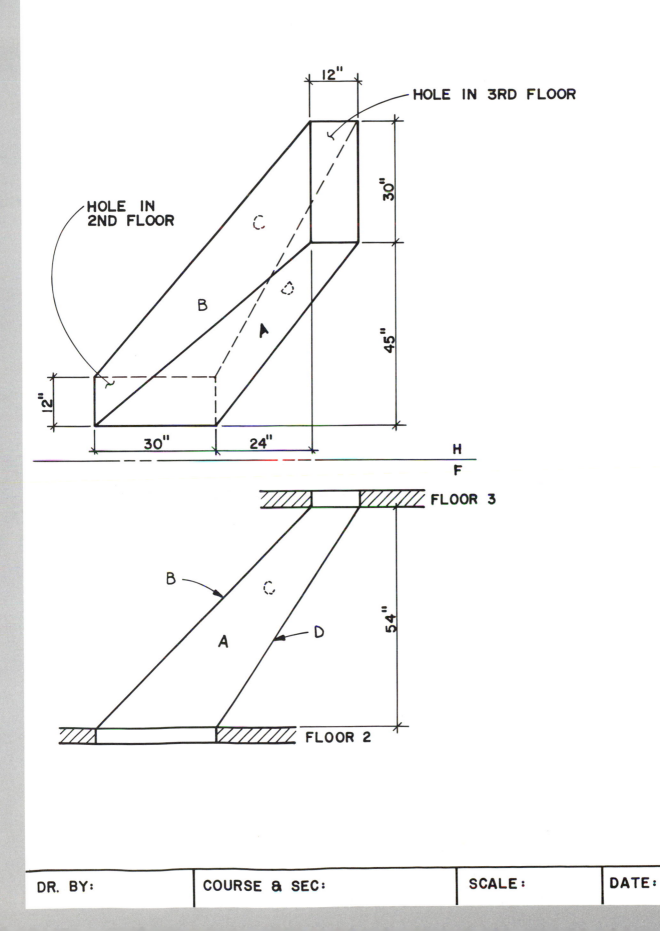

CHAPTER SEVEN
TEST

1. In determining the shortest distance from a point to a plane, how will the plane appear when the true length of the shortest distance is seen?

2. How would you locate the horizontal view of the shortest line from a point to a plane if the shortest line appears true length in a view adjacent to the horizontal view?

3. Define dihedral angle.

4. Describe the steps you would take to find the true angle between a line and a plane.

5. Explain the procedure for finding the true dihedral angle between two planes.

6. In your own words, explain why viewing the line of intersection between two planes as a point causes the planes to appear as edges.

7. Indicate whether the following statements are true or false.

    T F a. All shortest lines (horizontal, perpendicular, or grade) from a point to a plane have the same bearing.

    T F b. Two intersecting oblique planes will not appear as edges in the same view.

    T F c. The true angle between a line and a plane is seen when the plane is shown true shape.

    T F d. A view which has a line of sight parallel to the true length view of the line of intersection will show the true size of the dihedral angle.

    T F e. The true length of the shortest distance from a point to a plane is seen when the plane is in its true shape.

8. Explain in your own words why all shortest lines (horizontal, perpendicular, or grade) appear true length in the elevation view that shows the plane as an edge.

CHAPTER
EIGHT

Revolution

8.1 INTRODUCTION

There are two methods for solving many descriptive geometry problems. In the preceding chapters, all problems have been explained using the change of position method. When using this method, the drafter imagines that the object is in a fixed position. To obtain a different view, the observer moves to a new position from which to view the object. The object continues to remain stationary, as the observer moves around it. Generally, this method is more direct for many practical drawing problems, and is used almost unconsciously by most drafters.

The alternate method is revolution, which requires the observer to remain stationary and the object to be turned to obtain the various views. Some problems are solved more easily by revolution; therefore students should be familiar with it. In actual practice, a combination of the two methods described is used.

8.2 THE PRINCIPLES OF REVOLUTION

There are four fundamental principles that must be understood before you attempt to solve any problems by revolution. Studying these principles will teach you the concepts of what actually happens in space when revolution is used (see Figure 8-1).

1. When a point (**C**) is revolved in space, it is always revolved around a straight line used as an axis, (**AB**). It is important to know how the axis actually lies, before you attempt to revolve any point.
2. A point will revolve in a plane that is perpendicular to the axis, and its path is always a circle. The radius of the circle is the shortest distance from the point to the axis.
3. The circular path of the point is seen when the axis appears as a point, as seen in the horizontal view of Figure 8-1.
4. When the axis is shown true length, the circular path of the point will always appear as a straight line at right angles to the axis, as seen in the front view of Figure 8-1.

227

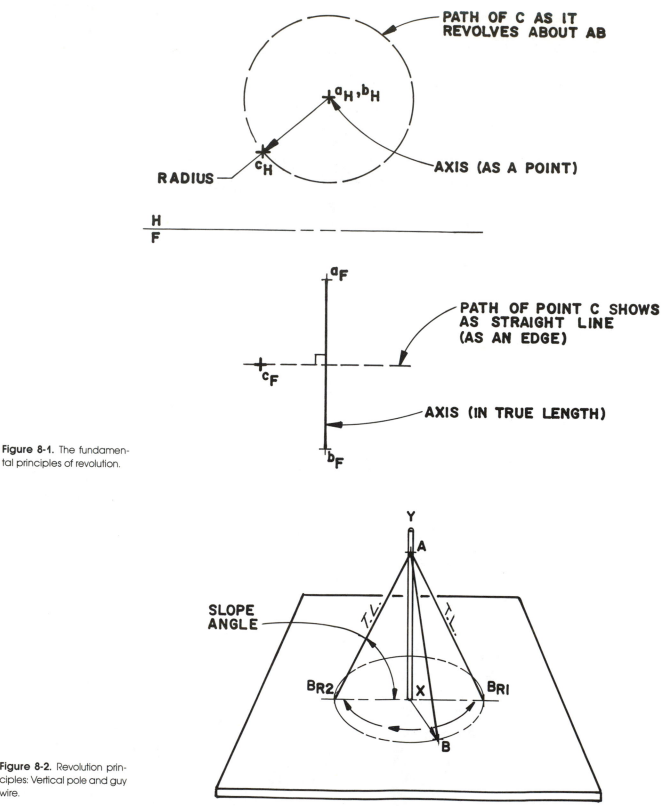

Figure 8-1. The fundamental principles of revolution.

Figure 8-2. Revolution principles: Vertical pole and guy wire.

8.3 THE TRUE LENGTH OF A LINE

When a line is parallel to a fold line in one view, it will appear true length in the adjacent view. When a line revolves about an axis, its position is changed, but its length remains the same. Examine Figure 8-2, which shows a vertical pole (axis) about which is rotated guy wire **AB**. You can see that guy wire **AB** is shown true length, when it is rotated into a position parallel with the front plane (**AB$_{R1}$** and **AB$_{R2}$**), and that its true slope also is shown in the rotated position.

Figure 8-3 shows the orthographic projection of the pole and the guy wire shown in Figure 8-2. In the horizontal view, the pole (axis) appears as a point. When the wire **AB** revolves about the pole, point **B** follows the indicated circular path. Point **A**, being directly on the axis, stays fixed. When b$_H$ is revolved to either b$_{HR1}$ or b$_{HR2}$, the guy wire is brought into a position parallel to fold line H/F. Because of this position, the adjacent (front) view of **AB** will be true length. In the front view, point b$_F$ moves perpendicular to the axis to either of the revolved positions, b$_{FR1}$ or b$_{FR2}$. Either position gives the desired true-length solution.

Figure 8-4 shows a line, **XY**, revolved about an axis that appears as a point in the front view. Point **Y** is revolved about the axis in the front view until x$_F$y$_{FR}$ is parallel to fold line H/F. The circular path of point **Y** is seen in the front view. In the horizontal view the movement of point y$_H$ is, as always, perpendicular to the axis, and in alignment with y$_{FR}$. Line x$_H$y$_{HR}$ is now true length.

For the purpose of finding the true length of a series of lines, the revolution method is usually the most efficient. It is used often in sheet metal work.

8.3.1 The Slope of a Line

The slope angle of a line will be seen in its true size, when the line is revolved about a vertical axis until the line appears true length in an elevation view. That the slope angle will be seen only in an elevation view is true for both the revolution method and the auxiliary-view method. Figure 8-5 shows again the vertical pole, **XY**, and guy wire **AB**.

When b$_H$ is rotated to b$_{HR1}$, parallel to fold line H/F, the true length of wire **AB** appears in the front view (a$_F$b$_{FR1}$). This is an elevation view and as such shows the true slope of the wire. If we revolved **AB** about a horizontal (level) axis, the true length appears in the horizontal view, as a$_H$b$_{HR2}$. But the true slope is **not** seen.

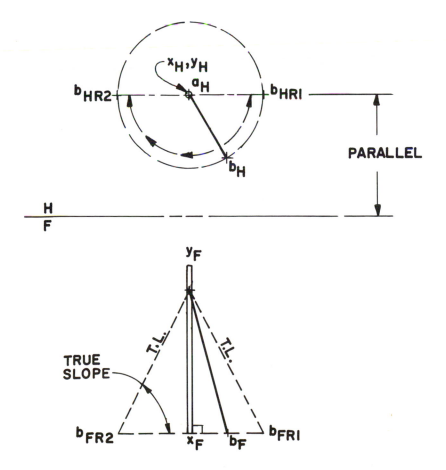

Figure 8-3. Orthographic projection: Vertical pole and guy wire.

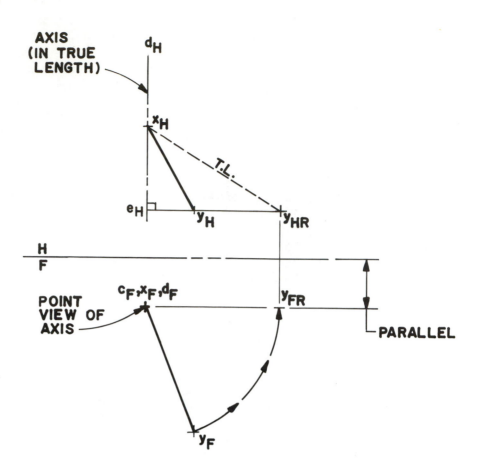

Figure 8-4. Revolution of a
line to obtain true length.

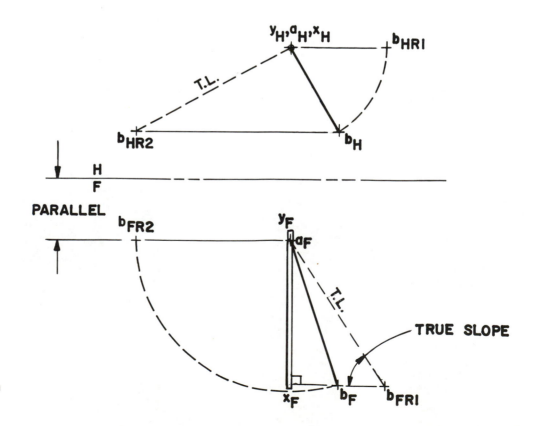

Figure 8-5. Revolution of a
line to obtain true slope.

8.4 THE TRUE SHAPE OF A PLANE

When a plane is seen true shape in a view, it *must* appear as an edge view, parallel to the fold line in all views adjacent to the true-shape view. In Figure 8-6, two views of a wedge block are shown. The problem is to revolve plane **ABC** until is appears true shape in the top view.

The edge view of the plane will be revolved about an axis that lies in the plane and that appears true length in the view where the true shape is to be shown. This axis of revolution is the horizontal line **AD**. View 1 is then drawn showing the axis **AD** as a point (a_1d_1) and plane **ABC** as an edge. The edge view is revolved about axis a_1d_1 until it is parallel to fold line H/1. Points b_1 and c_1 follow the circular paths shown to their new positions, b_{1R} and c_{1R}. In the adjacent horizontal view, points b_H and c_H *must move perpendicular* to the axis to their new positions b_{HR} and c_{HR}. Remember points a_H and d_H lie on the axis and do not move. The dashed plane $a_Hb_{HR}c_{HR}$ is the true shape required.

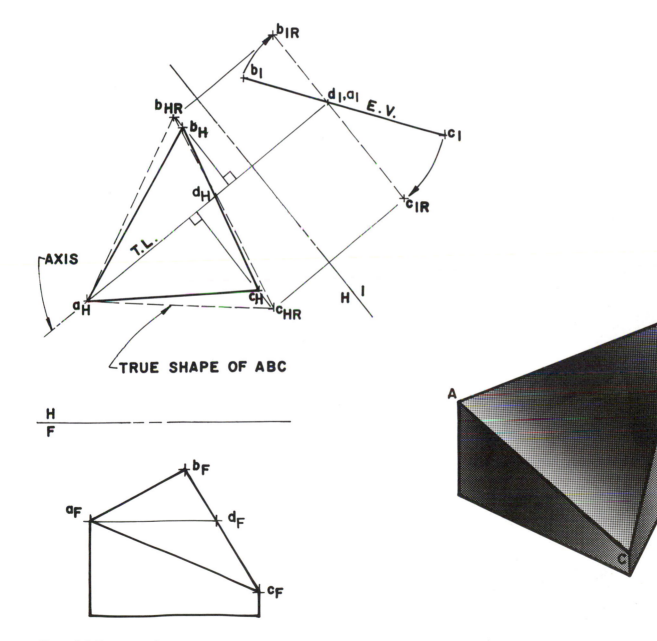

Figure 8-6. Revolution of a plane to obtain true shape.

It is not conventional to show the revolved plane in the front view. Notice that the chief accomplishment of revolution is to eliminate drawing one view. As you continue through the problems in this chapter, you should continue to see that the process of revolution is only a substitute for the final auxiliary view.

8.5 THE ANGLE BETWEEN A LINE AND A PLANE

The angle between a line and a plane may be seen when the line is revolved around an axis that is perpendicular to the plane. When the line is revolved to true length, the line-plane angle will be in its true size in the view where the plane appears as an edge. Figure 8-7 shows an A-frame, **XYZ**, and two guy wires, **AY** and **BY**, fastened to the frame at **Y**. It is necessary to find the true angle between line **AY** and the plane of the A-frame, **XYZ**.

To find the size of the angle, view 1 has been drawn to show plane **XYZ** as an edge, and view 2 to show the plane in true shape. The axis of revolution is a point in view 2, and is seen true length in view 1. In view 2, point **A** is revolved about the axis, until it is parallel to fold line 1/2. Projecting line **AY** back to view 1, it appears true length (y_1a_{1R}). Since the line now appears true length in the view which shows the plane as an edge, the true size of the angle between them may be seen.

8.6 THE ANGLE BETWEEN TWO PLANES (DIHEDRAL ANGLE)

The dihedral angle between two planes may be seen on a plane that is perpendicular to the line of intersection of two given planes. If we create a perpendicular cutting plane

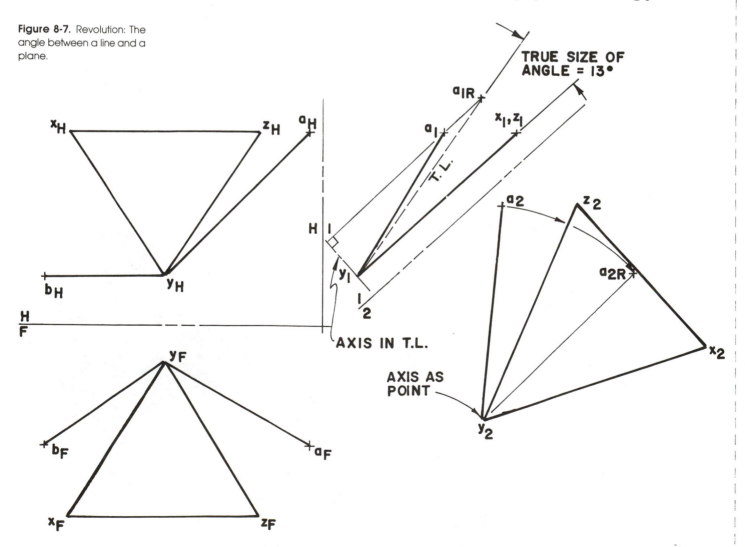

Figure 8-7. Revolution: The angle between a line and a plane.

and find its intersection with each plane, then revolution may be used to show the true dihedral angle. Figure 8-8 illustrates a hopper for which angles will be bent. It is necessary to find the dihedral angle between planes **ABCD** and **ABFE**. View 1 is constructed so the line of intersection, **AB**, is shown true length. Next, a cutting plane, **MN**, is drawn perpendicular to intersection line a_1b_1, and is assumed to be an edge in view 1. Although the cutting plane must be perpendicular to the line of intersection, it may intersect a_1b_1 at any desired location along its length. The cutting plane intersects plane **ABCD** along x_1y_1 and plane **ABFE** along x_1z_1. Project plane **XYZ** to the horizontal view. Now, revolve the cutting plane so that **XYZ** is drawn true shape in the top view. The true dihedral angle will be visible here. Again, notice in comparison to the auxiliary-view method, that the process of revolution simply eliminates the need for the final auxiliary view.

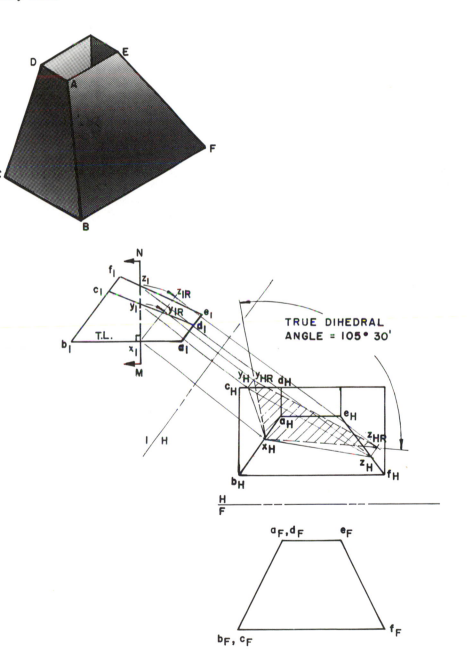

Figure 8-8. Revolution: The angle between two planes.

CHAPTER PROBLEMS

Problem 1: A radio transmission tower, **OP**, is supported by three guy wires, **OA**, **OB**, and **OC**. Find true length and true slope of all three guy wires, using the revolution method. Scale: $1'' = 60'$

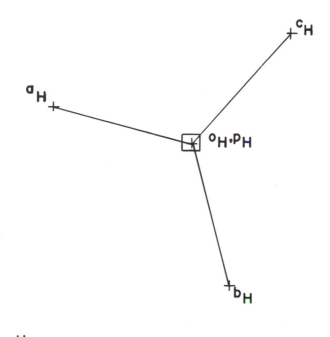

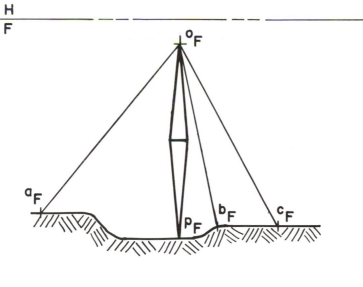

DR. BY:	COURSE & SEC:	SCALE:	DATE:

Problem 2: A derrick is located on a hillside having a 15 percent grade. The base of the mast is shown at p_F. The complete derrick consists of a central mast, **OP**, a boom, **MN**, and four guy wires. The guy wire directions are shown in the top view. The guy wires must clear the boom as it revolves at any angle with the mast. Locate the guy wire anchor points **A, B, C,** and **D** in both views, and determine the true length of each guy wire. Solve this problem using the revolution method. Scale: 3/32" = 1'-0"

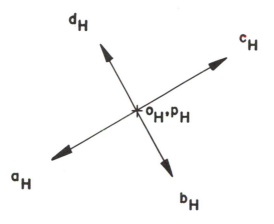

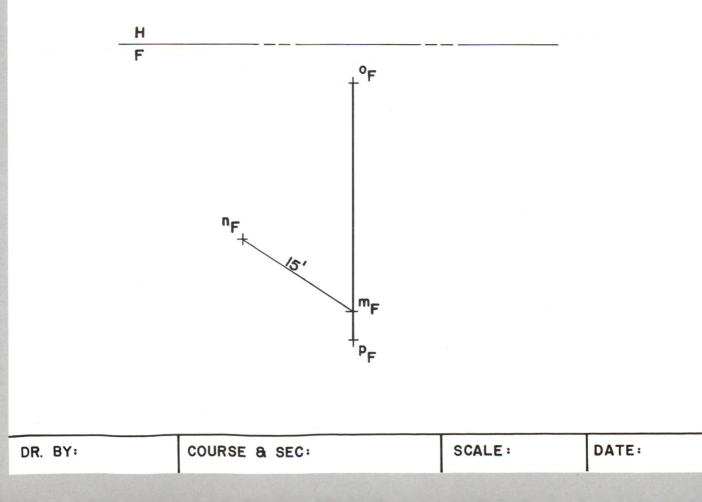

| DR. BY: | COURSE & SEC: | SCALE: | DATE: |

Problem 3: A length of 2-inch diameter steel tubing is bent as seen below. Determine the true length of the centerlines of each segment, **WX**, **XY**, and **YZ**, and indicate the slope of each segment. Determine the number of degrees in the bend at **X** and at **Y**. Use the revolution method. Scale: 1" = 4"

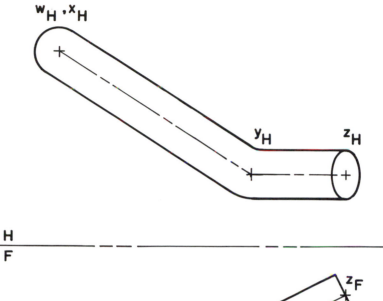

$$\frac{H}{F}$$

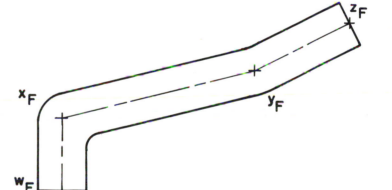

Problem 4: Surface **WXYZ** must be revolved into the horizontal position for milling purposes. Through what angle must the block be rotated about **WX** in order to bring surface **WXYZ** into the horizontal plane? Show the true shape of this surface in dashed lines on the horizontal view. Scale: Full

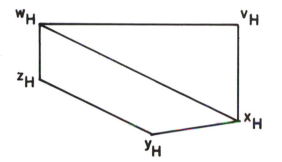

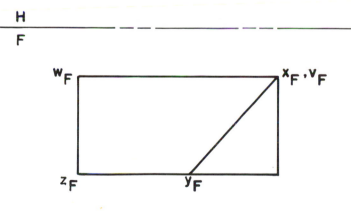

$\dfrac{H}{F}$

Problem 5: Two views of a special pipe fitting are shown. Find
the bend angle at **B** by the revolution method. Scale: Half

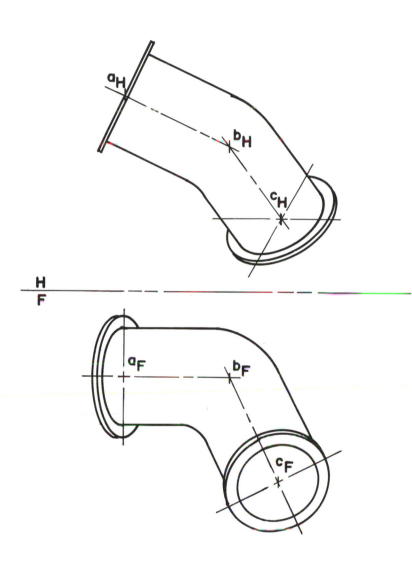

H
F

Problem 6: An X-frame stamping is illustrated. Find the hole center to center distances, **AB**, **BC**, **CD**, and **AD**, and the angle **CED**. Solve this problem by the revolution method. Scale: $\frac{1}{4}" = 1"$

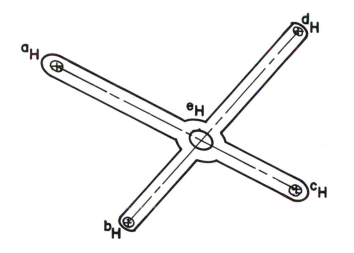

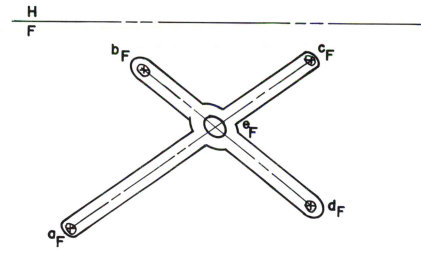

| DR. BY: | COURSE & SEC: | SCALE: | DATE: |

Problem 7: Line **XY** is the centerline of a ¼-inch diameter air-plane control cable that passes around a pulley at **X** and continues to **Y** through a 1-½-inch hole in the bulkhead. Find the angle that cable **XY** makes with the bulkhead surface, using the revolution method. Scale: ¼" = 1'-0"

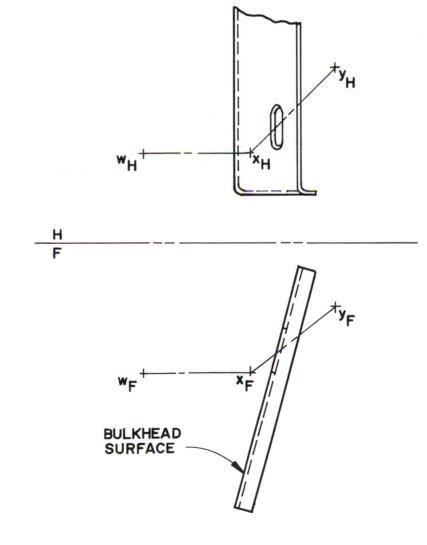

H
———
F

BULKHEAD
SURFACE

| DR. BY: | COURSE & SEC: | SCALE : | DATE : |

Problem 8: Line **XY** is the centerline of a hawsepipe (a pipe through the hull of a ship through which the anchor chain passes) connecting the hull and the deck of the ship. The deck and the hull are drawn as plane surfaces for simplicity. Using the revolution method, find the true length of the centerline **XY** and the angle it makes with the deck and the hull. Trace the problem, and complete your solution on a B-size drawing sheet. Scale: ⅛" = 1'-0"

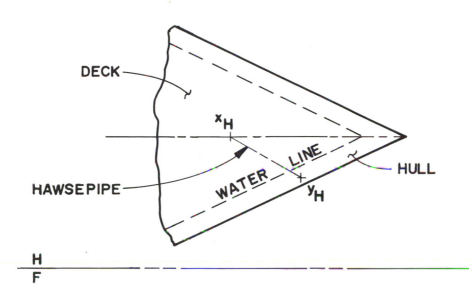

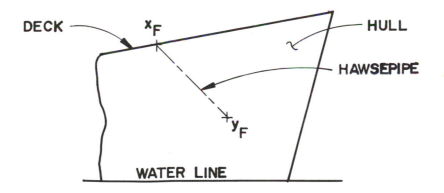

| DR. BY: | COURSE & SEC: | SCALE: | DATE: |

Problem 9: ABCDEF represents part of a bent floor plate. A column, **JK**, a cable, **MN**, and a brace, **GH**, pass through the floor. Using the revolution method, determine the angles that the column, the cable, and the brace make with the floor plate. Solve this problem on a C-size (17" × 22") drawing sheet using a scale of 1" = 4".

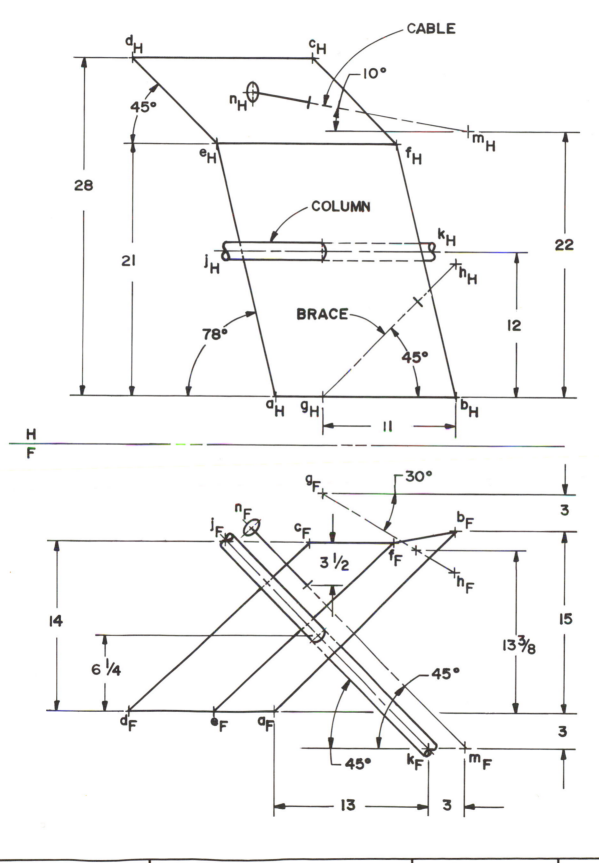

DR. BY: COURSE & SEC: SCALE: DATE:

Problem 10: Two views of a sluiceway are shown. Find the angle between the bottom surface, **A**, and the side, **B**, using the revolution method. Scale: 1" = 10'

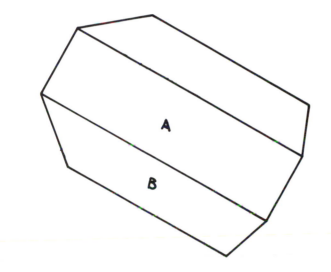

H
F

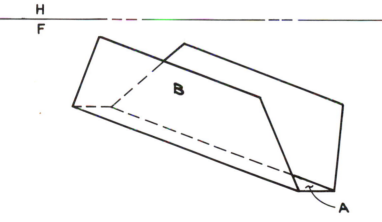

| DR. BY: | COURSE & SEC: | SCALE: | DATE: |

Problem 11: The hopper below is made of ¼-inch steel plate. The adjacent side plates of this hopper are to be riveted with a special corner angle placed inside the hopper as shown at $a_H e_H$. Using the revolution method, determine the bend angle for the corner angle **AE**, so that a detail drawing can be made for the shop. Do the same for corners **BF**, **DH**, and **CG**. Scale the drawing below and reproduce your solution on a B-size (11" × 17") drawing sheet. Scale: 3/16" = 1'-0"

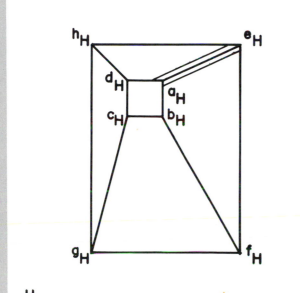

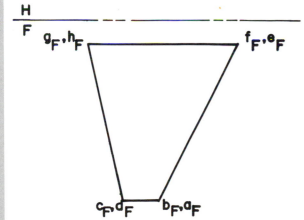

| DR. BY: | COURSE & SEC: | SCALE: | DATE: |

Problem 12: Two pieces of structural channel (6 × 2 × 8.2 lb/ft) are to be joined as shown. Find the dihedral angle between the two 6-inch surfaces using the revolution method. This prob- lem should be solved on a C-size (17" × 22") drawing sheet. (Note: A cross section of 6 × 2 channel is shown.) Scale: ¼" = 1"

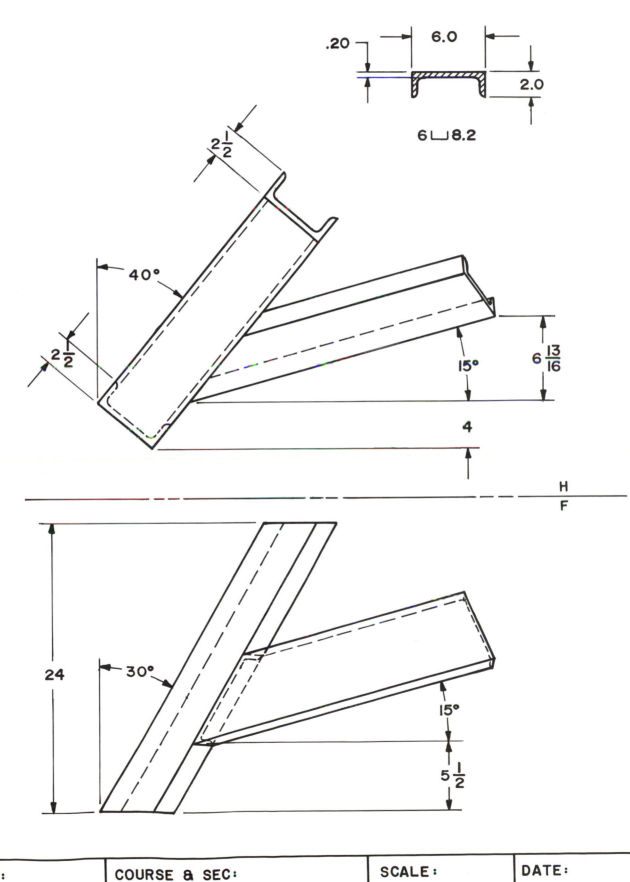

.20

6.0

2.0

6 ⊔ 8.2

2½

40°

2½

15°

$6\frac{13}{16}$

4

H
F

24

30°

15°

$5\frac{1}{2}$

| DR. BY: | COURSE & SEC: | SCALE: | DATE: |

CHAPTER EIGHT
TEST

1. Explain the difference between the change-of-position method and the revolution method of solving descriptive geometry problems.

2. List the four fundamental principles of revolution.

3. Explain how you would find the true length of a line by the revolution method. You may draw an example that will illustrate your explanation.

4. When finding the slope or grade of a line using the revolution method, three conditions must be met. List them.

5. Explain how you would find the true shape of a plane by revolution.

6. How must a plane appear in any view adjacent to its true shape view?

7. If a plane is to be revolved into true shape in a horizontal view, how will the axis of rotation appear in the horizontal view?

8. What conditions must be met to show the true angle between a line and a plane using the rotation method?

9a. Explain in your own words how you find the dihedral angle between two planes using the revolution method.

9b. How does this method differ from the auxiliary-view method studied in Chapter 7?

10. In general, what is accomplished by the process of revolution as compared to the auxiliary-view method?

CHAPTER
NINE

Piercing Points and the Intersections of Planes

9.1 PIERCING POINT OF A LINE AND A PLANE

If a line neither lies in a particular plane nor lies parallel to it, the line will intersect the plane at a single point. This point of intersection, called a **piercing point**, must lie in the plane and on the line. The problem of finding the piercing point of a line in a plane is the basic element in so many descriptive geometry problems that it could well be considered as important as the four fundamental views.

9.1.1 The Piercing Point of a Line and a Plane—Auxiliary-View Method

This method, also called the edge-view method, requires that the plane appear as an edge. **The piercing point is a point common to both the line the plane**. Since the edge view of the plane contains all points on the plane, the view that shows the edge view of the plane will also show the point where the line pierces the plane.

The simplest case occurs when the given plane appears as an edge in one of the given views. Figure 9-1 shows the pictorial and orthographic views of a pipe, **MN**, which intersects a wall, **ABCD**. In the orthographic drawing, the top view shows the edge view of the wall and the centerline of pipe, **MN**. The intersection of the pipe with the wall is determined at point p_H, which is common to both. This point is projected to the front view at p_F. The visibility is determined by simple observation. You can see that $p_H m_H$ is behind the wall, as you view from the top, therefore $p_F m_F$ is hidden in the front view.

A portion of an aircraft bulkhead, **ABCD**, is intersected by cable, **XY**, in Figure 9-2. By finding the bulkhead as an edge in auxiliary view 1, the piercing point is identified (p_1) and can be projected to the horizontal view at p_H and front view at p_F. Visibility may be determined by the method explained in Chapter 6.

9.1.2 The Piercing Point of a Line and a Plane—Cutting-Plane Method

It is possible to find the piercing point of a line and a plane by using only two views and a cutting plane. This is an important and frequently used technique based on the following principle: **the intersection of a line with a plane will lie on the intersection of a**

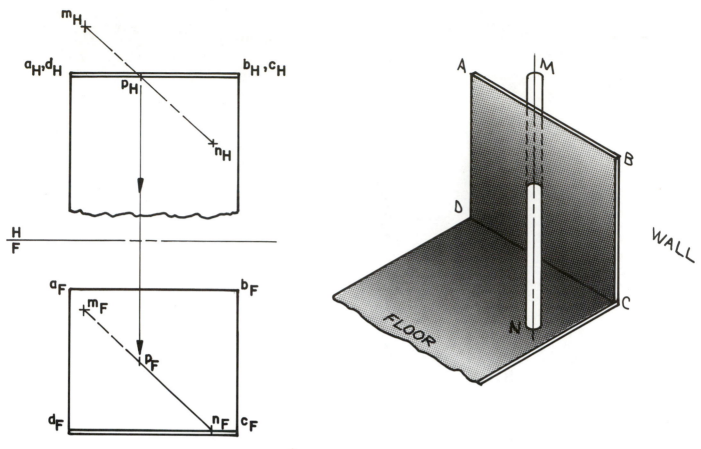

Figure 9-1. Piercing point of a line and a plane: Edge-view method.

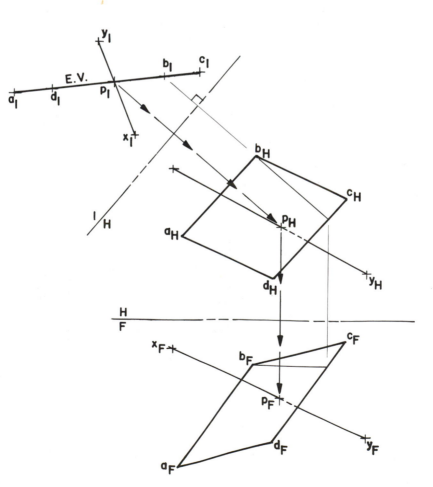

Figure 9-2. Piercing point of a line and a plane: Edge-view method.

given plane and a cutting plane that contains the line. Figure 9-3 shows the pictorial and the orthographic views of an oblique plane **ABC** and a line **XY**. First let us examine the pictorial drawing. A vertical cutting plane has been placed so that line **XY** lies entirely within it. The vertical cutting plane and plane **ABC** must intersect on a straight line. This line is labeled **DE**. The given line **XY** and the intersection line **DE** both lie in the vertical cutting plane, and must be either parallel or intersecting. In this case, they intersect at **P**, which is the desired intersection between line **XY** and plane **ABC**.

Let us now apply this to the orthographic drawing. The vertical cutting plane is shown in the top view, where it appears as an edge view and contains line **XY**. This cutting plane could be extended indefinitely but is drawn only as long as necessary. The next step is to find the line of intersection, **DE**. Note that line **AC** intersects the cutting plane at point d_H and line **BC** intersects the cutting plane at point e_H. These points are projected to the front view and labeled $d_F e_F$. When connected, line $d_F e_F$ is the front view of the line of intersection, which intersects $x_F y_F$ at p_F. This is the required piercing point line **XY** makes with plane **ABC**. Point p_H is located on line $x_H y_H$ by simple projection.

The cutting plane may also be selected so it appears as an edge in the front view.

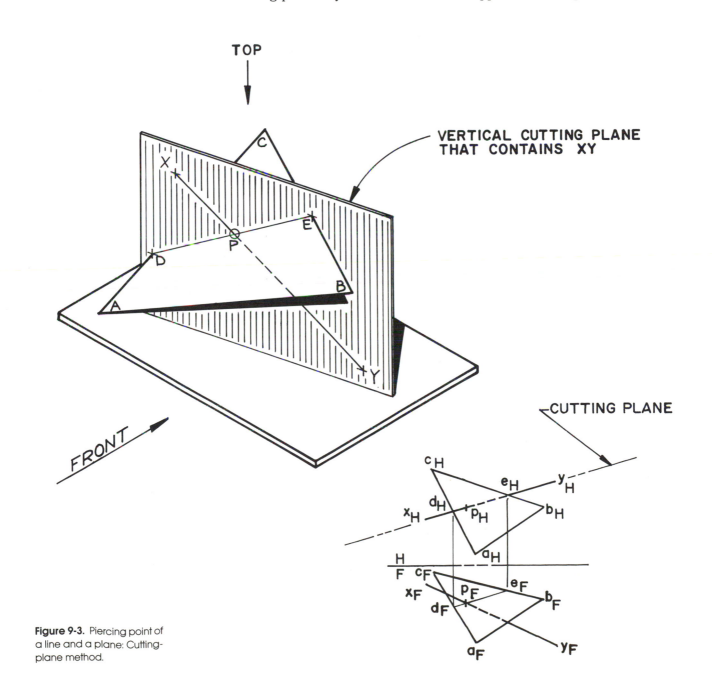

Figure 9-3. Piercing point of a line and a plane: Cutting-plane method.

Figure 9-4 illustrates, pictorially and orthographically, just such a situation. In the pictorial representation, a cutting plane is placed so as to contain the line **MN** and intersect the plane **ABC**. The line of intersection formed by the cutting plane and plane **ABC** is labeled **GH**. Again, point **P** is the required piercing point. In the orthographic drawing, the cutting plane is made to appear as an edge in the front view, and contains line **MN** ($m_F n_F$). The line of intersection $g_F h_F$ is projected to the horizontal view, $g_H h_H$. Line of intersection $g_H h_H$ crosses line $m_H n_H$ at their common point, the piercing point, p_H. The front view of the piercing point, p_F, is found by projection.

If it should happen that the given line is a profile line, the cutting-plane method will not produce a solution in the horizontal and front views only. This is because the cutting plane appears as an edge in both views. Since a new view must be drawn, it is usually preferable to use the edge-view method for such a case.

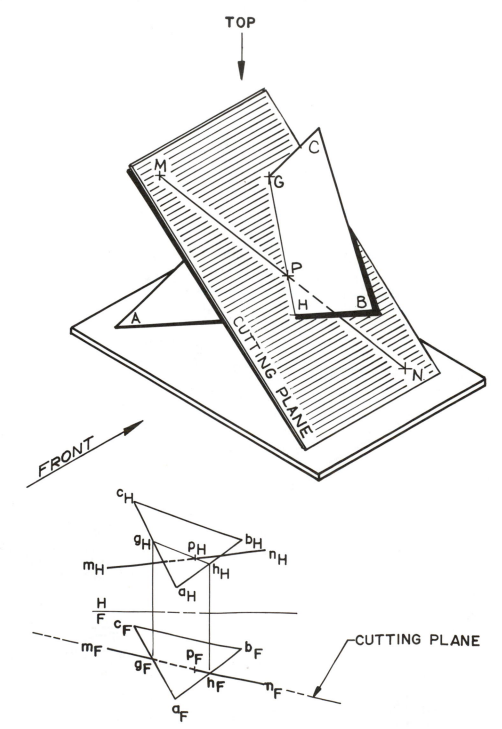

Figure 9-4. Piercing point of a line and a plane: Cutting-plane method.

9.2 THE INTERSECTION OF TWO PLANES

If two planes are not parallel, they will intersect along a straight line that is common to both planes. Because the position of this shared line may be fixed by any two points on the line, you need only to find two points that lie on both planes to find the intersection.

9.2.1 The Intersection of Two Planes—Edge-View Method

If one of the given planes appears as an edge in a given view, then a part of the solution is already available. Figure 9-5 illustrates this type of problem. In the horizontal view, plane **ABCD** and plane **XYZ** intersect. It is evident that the line **XZ** intersects plane **ABCD** at point **R**, and line **XY** intersects plane **ABCD** at point **S**. Points **R** and **S** lie in plane **XYZ** and also in plane **ABCD**, and are, therefore, two points on the required line of intersection. Points r_F and s_F are located in the front view by projection. Visibility is then determined.

When both of the planes are oblique as seen in Figure 9-6, the solution may be found by showing one of the planes as an edge. Either plane may be selected for projection to an edge view. In Figure 9-6, plane **RST** is found as an edge in view 1. Plane **XYZ** is also shown in view 1. The intersection points of lines **XY** and **XZ** are shown at a_1 and b_1, respectively. These two points can be projected to the horizontal view to located a_H and b_H. Remember that they lie on lines **XY** and **YZ**. The line of intersection, $a_F b_F$, is similarly located in the front view, and visibility is determined.

9.2.2 The Intersection of Two Planes—Cutting-Plane Method

It is important to understand that when you are finding the line of intersection between two planes, it is simply a matter of twice finding the piercing point of a line and a plane, thus locating two piercing points which form the line of intersection.

In Figure 9-7 planes **XYZ** and **RST** are intersecting. Their line of intersection was found by using the two cutting planes to locate two piercing points. Line **XY** is arbitrarily chosen first, and its intersection with plane **RST** is found by using the vertical cutting plane, **CP-1**. This cutting plane intersects plane **RST** along line **AB**, and in the front view $a_F b_F$ intersects $x_F y_F$ at p_F. This point, **P**, is where line **XY** intersects plane **RST** and is one point on the required line of intersection.

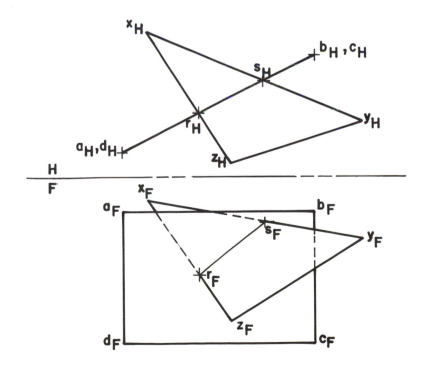

Figure 9-5. Intersection of two planes: Edge-view method.

A second vertical cutting plane, **CP-2**, containing line **YZ**, is now selected. Line **YZ** intersects plane **RST** along line **CD**, and in the front view $c_F d_F$ intersects $y_F z_F$ at point q_F. This point **Q**, where line **YZ** intersects plane **RST**, is the second point on the required line of intersection. The horizontal view of the line of intersection, $p_H q_H$, is found by projection. Visibility is determined.

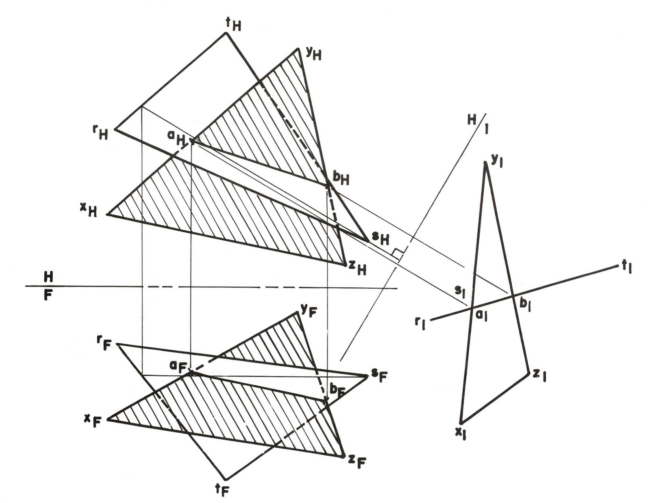

Figure 9-6. Intersection of two planes: Edge-view method.

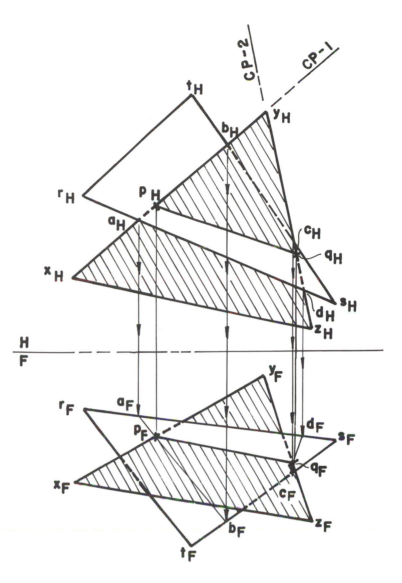

Figure 9-7. Intersection of two planes: Cutting-plane method.

CHAPTER PROBLEMS

Problem 1: Two views of a rectangular-shaped bin are shown. It has a sloping bottom, **WXYZ**, which is intersected by a 2-inch diameter pipe, **CD**. The centerline of the pipe is given. Find where the pipe pierces the bottom of the bin by using the edge-view method. Show the complete pipe, and its correct visibility in the horizontal and front views. Scale: 1" = 5'

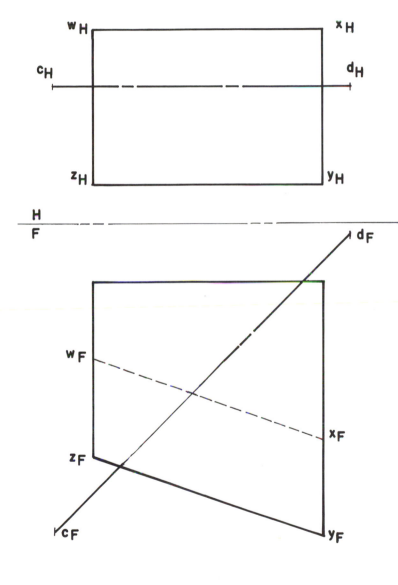

DR. BY:	COURSE & SEC:	SCALE:	DATE:
			1" = 5'

Problem 2: Find the points at which the centerline, **XY**, of a culvert intersects the fill embankments having a 1 to 1 horizon- tal to vertical grade. Use the cutting-plane method, and check your solution with the edge-view method. Scale: 1" = 100'

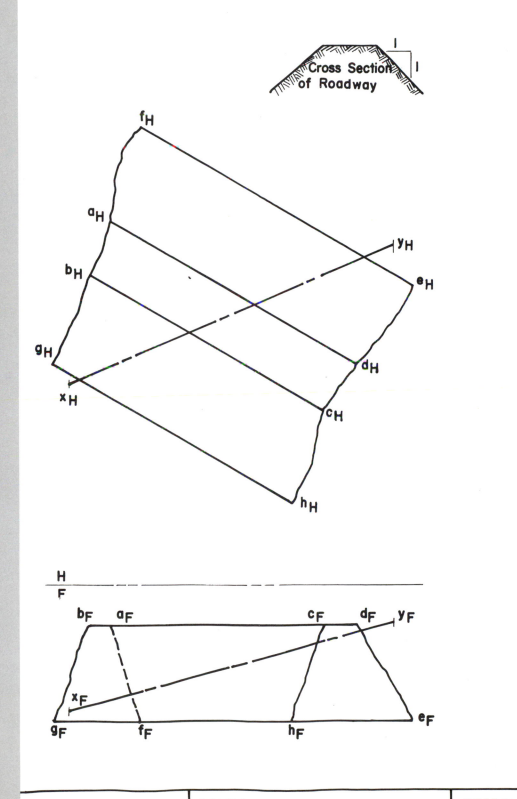

Cross Section of Roadway

Problem 3: A lower plate, **WXYZ**, is installed in a test tunnel. An opening must be cut in the plate to allow for the passage of an accelerating particle moving in the direction shown by the arrow. Use the cutting-plane method to find the center of the opening. Show the correct visibility. Scale: 1" = 20'

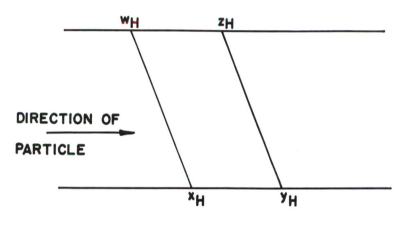

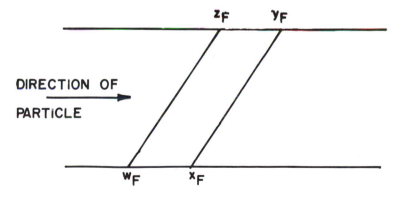

| DR. BY: | COURSE & SEC: | SCALE: | DATE: |

Problem 4: A vein of ore has been determined by plane **RST**. A mine tunnel, **AB**, is dug toward the vein. How much must the tunnel be extended from point **A** to reach the ore vein? Scale: 1" = 200'

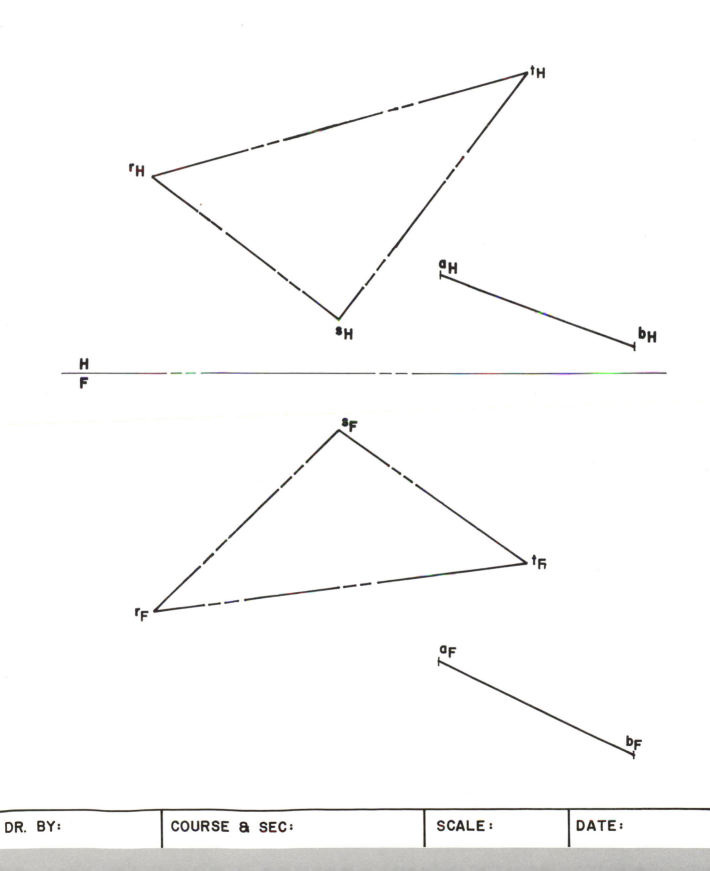

DR. BY:	COURSE & SEC:		SCALE:	DATE:

Problem 5: A rectangular frame, **WXYZ**, is supported by two wooden braces, **RS** and **TU**. Find the points where the braces pierce the plane of the frame. How much of each brace must be removed if they are not to extend beyond the plane of the frame? Scale: ¼" = 1'-0"

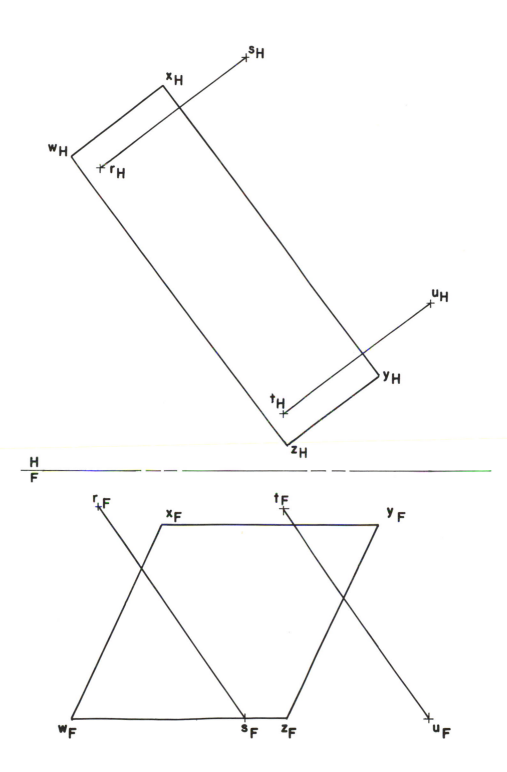

DR. BY: COURSE & SEC: SCALE: DATE:

Problem 6: Several contour lines on a map are shown as parallel straight lines for simplicity. Assume that the ground is composed of planes between any two contour lines. From a point **X** on the 1525-foot contour line, the centerline of a road bears S30°W, and has a falling grade of 10 percent. It will cross the valley on fill and will tunnel through the hill. Show the centerline of the road in the two given views. Using the cutting-plane method, find the points where the centerline of the road enters the tunnel, and where it comes out again. Draw an edge view of the ground surface and the true distance from point **X** to the tunnel opening. Scale the drawing below and reproduce your solution on a B-size (11" × 17") drawing sheet. Scale: 1" = 100'

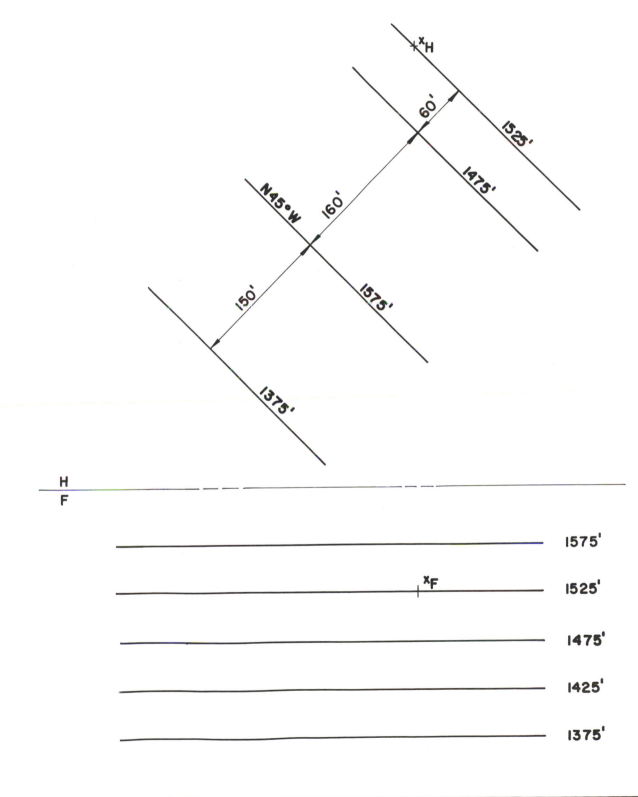

DR. BY:	COURSE & SEC:	SCALE:	DATE:

Problem 7: Locate, label, and give the bearing of the line of intersection, **XY**, between the two given planes, **RST** and **ABC**. Use the cutting-plane method, and check your solution with the edge-view method.

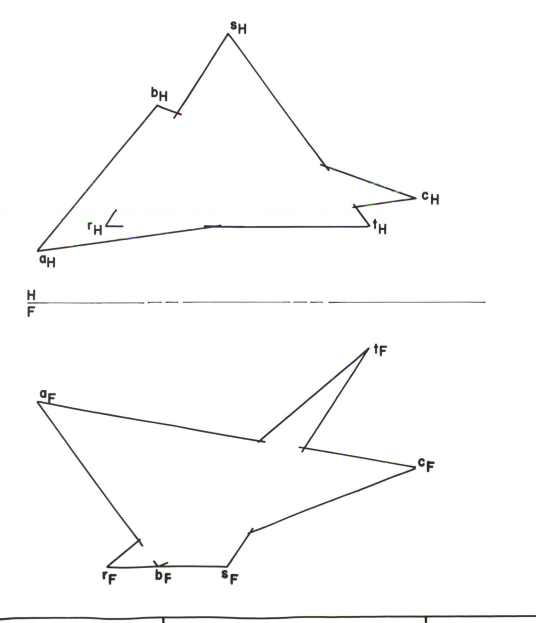

| DR. BY: | COURSE & SEC: | SCALE : | DATE : |

Problem 8: Locate, label, and give the bearing of the line of intersection between planes **RST** and **WXYZ**, utilizing the cutting-plane method. Check your solution with the edge-view method. Show the correct visibility.

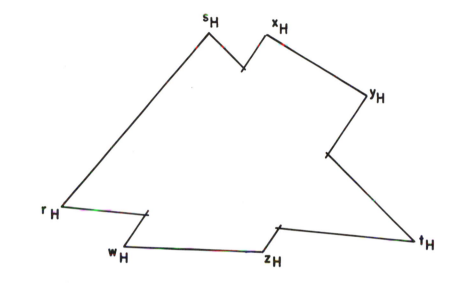

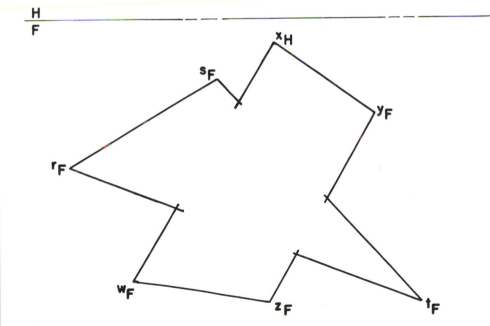

| DR. BY: | COURSE & SEC: | SCALE: | DATE: |

Problem 9: Two views of intersecting triangular prisms are shown. Complete the front view showing the lines of intersection. Allow your construction to show. Include the correct visibility.

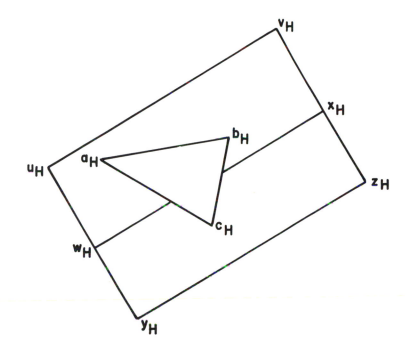

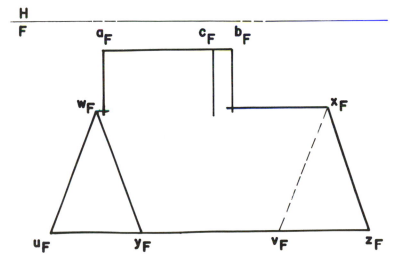

| DR. BY: | COURSE & SEC: | SCALE: | DATE: |

Problem 10: Two views of a house and a detached garage are shown. The house is to be extended to the garage, without changing the existing roof slopes of both buildings. Find the intersections of the extended roof of the house with the garage. Use the cutting-plane method to find the intersections. Include the correct visibility in both views. Scale: $\frac{1}{8}" = 1'\text{-}0"$

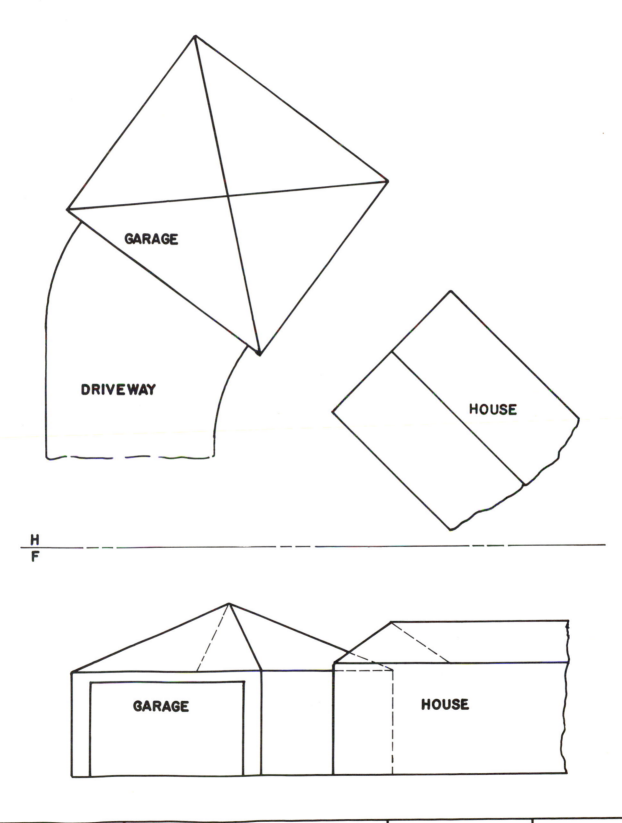

GARAGE

DRIVEWAY

HOUSE

H
F

GARAGE

HOUSE

| DR. BY: | COURSE & SEC: | SCALE: | DATE: |

CHAPTER NINE

TEST

1. Define a piercing point.

2. What are the two methods used to find the point of intersection of a line and a plane?

3. Explain why finding the edge view of a plane will show the point where the line intersects the plane.

4. When determining the point where a line pierces a plane, could the point of intersection be identified in a view where the line appears as a point? Explain your answer.

5. Which method of finding the intersection of a point and an oblique plane requires only two views?

6. To determine the piercing point of a line and an oblique plane, an edge view of a cutting plane can be used. Provided that the horizontal and front views of the given line and plane are shown, in which view can the cutting plane be indicated?

7. Explain the cutting-plane method of finding the point of intersection between a line and a plane.

8. The intersection between two nonparallel planes will take what form?

9. How many points must you find to determine the intersection between two non-parallel planes?

10. Explain how you would find the line of intersection of two planes using the edge-view method. You may use a drawing to illustrate your explanation.

11. Explain how you would find the line of intersection between two nonparallel planes using the cutting-plane method. You may use a drawing to illustrate your explanation.

CHAPTER
TEN

Developments

10.1 INTRODUCTION

In addition to finding the line of intersection between various surfaces and geometric solids, you must be able to reproduce many of these solids on a flat surface such as sheet metal or sheet plastic. When the geometric shape is laid out in a flat pattern, certain sides are connected so that the item can be rolled or bent into the original shape of the solid. The solid shape that is unfolded or unrolled onto a flat plane is known as a **development**. **The important characteristic of these surface developments is that all lines appear true length in the development**. Figure 10-1 pictorially illustrates a simple development of a right, rectangular prism that can be assumed to be made of sheet metal. It shows the faces of the prism being unfolded sequentially to form the flat pattern shown at (d).

Many manufactured articles are fabricated from sheet metal by cutting and bending the material into the desired shapes. A development of the surfaces of the object is made first, either on paper or directly on the flat surface of the metal. If large quantities of the part are required, a metal pattern, or template, of the development may be made. The outline of the pattern may then be transferred to the flat stock. The **bend lines** are often located by means of small punch marks. After the metal is cut to the desired shape, it is bent, curved, or pressed into its finished form. The edges are commonly joined by soldering, welding, riveting, or seaming. Additional metal is allowed for these joints. To account for the thickness of the metal, **bend allowances** are made for bending metals thicker than 24 gage (.0250 in.). All developments in this chapter will be theoretical, in that we will not allow for these slight additions and modifications necessary for joining and bending.

Developments are divided into three groups, according to the type of surface and the method of development used. Parallel-line developments are those found for prisms and cylinders. Radial-line developments are those obtained for pyramids and cones. Finally, there are triangulation developments, which are found by dividing a given surface into a series of triangular areas.

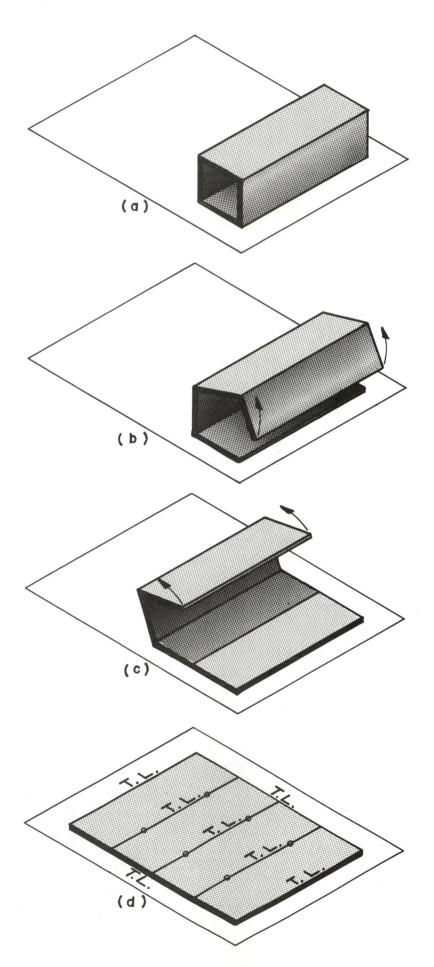

Figure 10-1. Development of a right rectangular prism.

10.2 PARALLEL-LINE DEVELOPMENTS

Parallel-line development is used for prisms and cylinders, where the lateral edges, or elements, are parallel to each other.

10.2.1 Development of a Right Rectangular Prism

In Figure 10-2, two views of a right rectangular prism are shown. Notice that the lateral edges appear true length in the front view, and that the perimeter, the actual distance around the prism, appears true length in the horizontal view. Note also that numbers are used instead of letters to label the development. This is customary for simplicity.

Figure 10-3 shows the development of the prism shown in Figure 10-2. The procedure for obtaining this development is as follows:

1. Construct a stretch-out line on which the perimeter of the prism will be unfolded.
2. Transfer the perimeter distances, 1 to 2, 2 to 3, 3 to 4, and 4 to 1 found in the horizontal view, to the stretch-out line. Identify points 1, 2, 3, 4, & 1 on the stretch-out line.
3. At each numbered point construct perpendicular lines.
4. Transfer the true length of the lateral edges, 1-1', 2-2', 3-3', and 4-4', found in the front view to their respective locations on the stretch-out line.
5. Connect points 1', 2', 3', 4', & 1' with a straight line to complete the development.

10.2.2 Development of a Truncated Right Prism

The horizontal, front, and auxiliary views of a truncated right prism are shown in Figure 10-4. By definition, the top of the truncated prism is not parallel to its bottom. In this example, the lateral surfaces are developed, with the top and bottom surfaces attached.

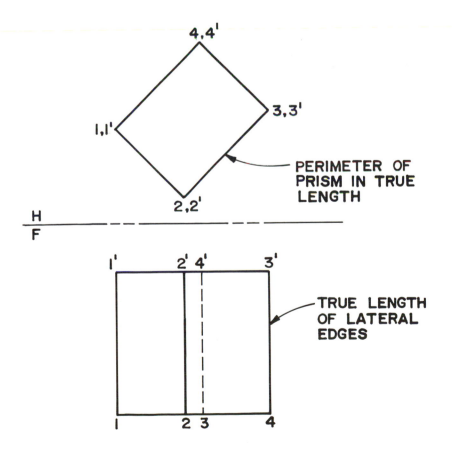

Figure 10-2. Two views of a right rectangular prism.

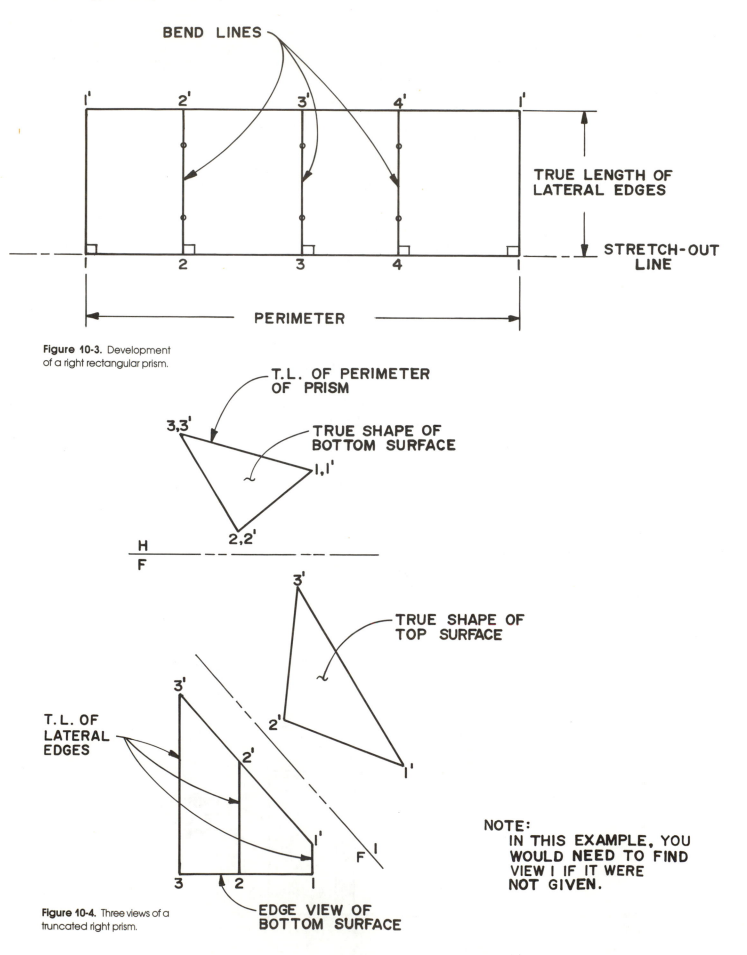

Figure 10-3. Development of a right rectangular prism.

Figure 10-4. Three views of a truncated right prism.

BEND LINES

1'   2'   3'   4'   1'

TRUE LENGTH OF
LATERAL EDGES

STRETCH-OUT
LINE

1   2   3   4   1

PERIMETER

T.L. OF PERIMETER
OF PRISM

3,3'

TRUE SHAPE OF
BOTTOM SURFACE

1,1'

2,2'

H
F

3'

TRUE SHAPE OF
TOP SURFACE

2'

1'

T.L. OF
LATERAL
EDGES

3'

2'

1'

F¹

3   2   1

EDGE VIEW OF
BOTTOM SURFACE

NOTE:
IN THIS EXAMPLE, YOU
WOULD NEED TO FIND
VIEW I IF IT WERE
NOT GIVEN.

Figure 10-5 shows the development of the truncated right prism shown in Figure 10-4. The procedure for finding this development is as follows:

1. Draw a stretch-out line and transfer the perimeter distance from around the prism to this line. Note points 1, 2, 3, and 1.
2. Construct lines perpendicular to the stretch-out line at points 1, 2, 3, and 1.
3. Begin with the shortest true length lateral edge, 1-1' and lay it out on the appropriate perpendicular line. Do the same with the other true length lateral edges, 2-2', 3-3', and then 1-1' again.
4. Connect 1' to 2', 2' to 3' and 3' to 1'.
5. Attach the true shape of the bottom surface, by transferring the true distances from the horizontal view. Then attach the true shape of the top surface, by transferring the true length distances from view 1.

Pay particular attention to the fact that this development began at the shortest lateral edge. In actual practice, it is most economical to join the desired form along its shortest edge, since it would require the least extra material, welding, riveting, etc.

10.2.3 Development of an Oblique Prism

Figure 10-6 shows a diagonal offset connector used to connect two rectangular ventilating ducts. Four views, a front, horizontal, primary, and secondary auxiliary are shown. View 1 is constructed to show the true lengths of the lateral edges, 1'-1", 2'-2", 3'-3", and 4'-4". View 2 is an end view (a right section) which shows the lateral edges as points and shows the true distance between the edges (the perimeter of the prism).

Figure 10-7 shows the development of this connector (an oblique prism). The procedure for constructing the development is as follows:

1. Note that, if not already given, you must construct a view of the prism in which the lateral edges appear true length, and a view in which you can see the perimeter of the prism.

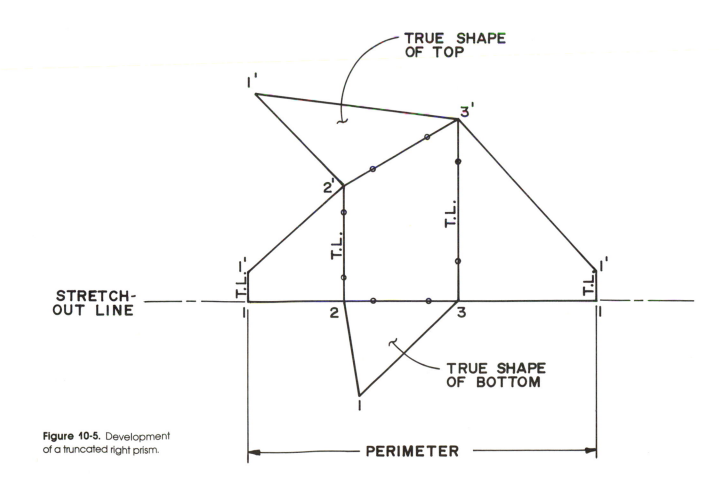

Figure 10-5. Development of a truncated right prism.

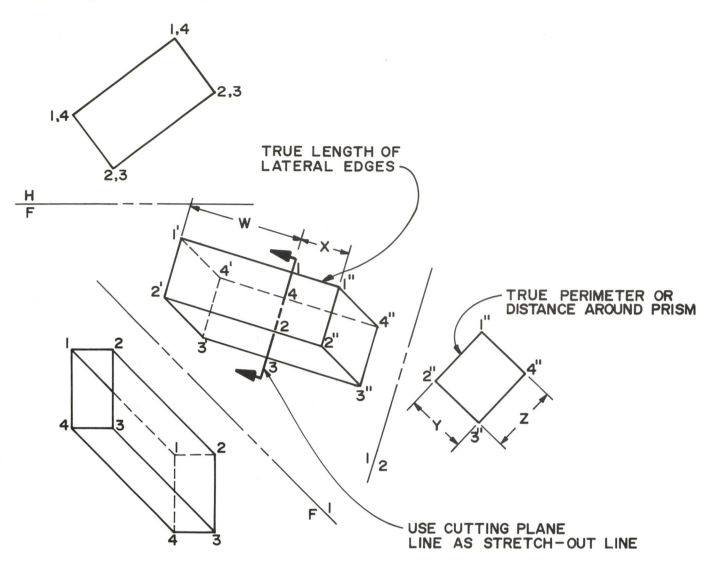

TRUE LENGTH OF LATERAL EDGES

TRUE PERIMETER OR DISTANCE AROUND PRISM

USE CUTTING PLANE LINE AS STRETCH-OUT LINE

Figure 10-6. Four views of an oblique prism.

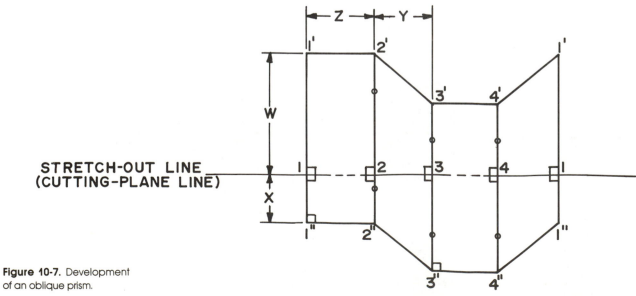

STRETCH-OUT LINE (CUTTING-PLANE LINE)

Figure 10-7. Development of an oblique prism.

2. Note also that in view 1 in Figure 10-6, a cutting plane is drawn perpendicular to the true length lateral edges of the prism. This cutting plane cuts the lateral edges at points 1, 2, 3, and 4.
3. Draw a stretch-out line and transfer the perimeter distances found in view 2. These are represented as distances **Y** and **Z**.
4. Construct perpendiculars to the stretch-out line at points 1, 2, 3, 4, and 1.
5. Starting with lateral edge 1'-1", transfer the true length of these edges to the development. Begin by transferring distance **W** above the stretch-out line and then distance **X** below it to locate edge 1'-1". Repeat this process to locate edges 2'-2", 3'-3", 4'-4", and 1'-1" respectively.
6. Connect points 1', 2', 3', 4' and 1' with straight lines. Do the same for points 1", 2", 3", 4" and 1" to complete the development.

10.2.4 Development of a Right Cylinder

The development of a cylinder is similar to the development of a prism. The cylinder is considered to be a prism with an infinite number of faces. But in practical applications, the number of faces are considered finite. The elements of the cylindrical surface are the same as the lateral edges of the prism. They are parallel, and must be viewed in their true lengths in order to use them in a development. As with the prism, the perimeter is seen in the end view, and is equal to the circumference (πD) of the circle. Figure 10-8 illustrates pictorially the development of a cylinder.

Two views of the cylinder are shown along with the development in Figure 10-9. The procedure for drawing the development is as follows:

1. Divide the circumference, seen in the horizontal view of the given cylinder, into a convenient number of radial divisions. For clarity, number these divisions. In this example, twelve divisions at 30° increments are used.
2. Project points 1 through 12 to the base of the cylinder in the front view. At these points construct straight-line elements on the cylinder surface. **Since the axis of the cylinder is true length in the front view, each element appears true length also.**

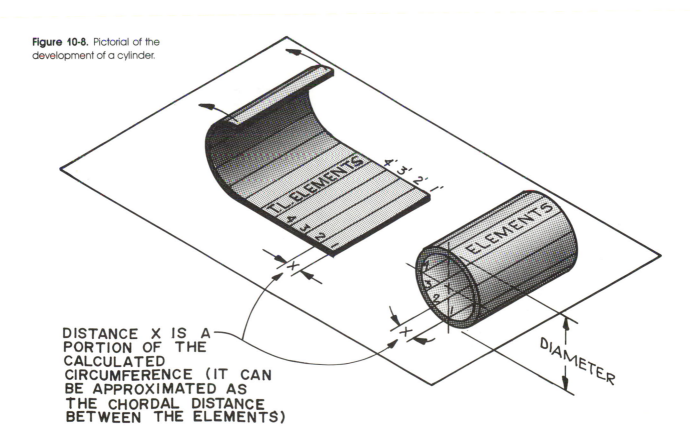

Figure 10-8. Pictorial of the development of a cylinder.

DISTANCE X IS A PORTION OF THE CALCULATED CIRCUMFERENCE (IT CAN BE APPROXIMATED AS THE CHORDAL DISTANCE BETWEEN THE ELEMENTS)

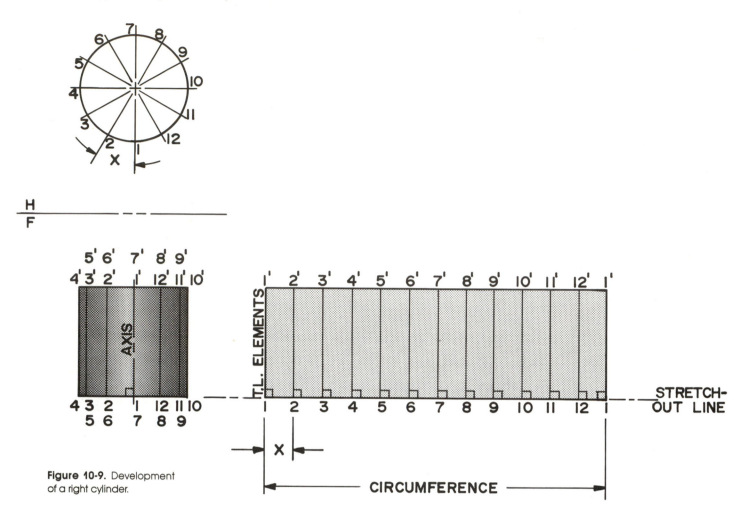

Figure 10-9. Development of a right cylinder.

3. Draw a stretch-out line equal in length to the circumference of the cylinder. The actual length of the circumference, seen in the top view, can be approximated by measuring the chord distance, **X**, and transferring it to the stretch-out line.
4. At each numbered point on the stretch-out line construct a perpendicular. The length of these lines will be equal to the true-length elements of the cylinder.
5. Connect points 1′ through 12′ and 1′ with a straight line to complete the development.

10.2.5 Development of a Truncated Right Cylinder

Like the truncated right prism, the truncated right cylinder has a top surface that is not parallel to its bottom. Because of this, the true lengths of the elements on the cylindrical surface will vary. Figure 10-10 shows the orthographic drawing and the surface development, including the top and bottom surfaces, of a truncated right cylinder.

The procedure for finding this surface development is as follows:

1. In the top view, on the circumference, divide the circle into equal radial divisions (12, in this example). Number the divisions 1 through 12.
2. Project these points to the bottom in the front view. Because the cylinder axis appears true length in the front view, these numbered elements will also appear true length. Label the elements 1-1′, 2-2′, etc., as shown.
3. Draw a stretch-out line equal in length to the circumference of the cylinder. Again you can find the actual length of the circumference in the horizontal view, or you can approximate it by transferring the chord distance, **X**, to the stretch-out line repeatedly.
4. Once you have transferred the circumference to the stretch-out line, number the points. Draw perpendicular lines at each point beginning on the shortest element at 1. The lengths of the perpendiculars will be equal to the true-length elements of the cylinder as seen in the front view. This locates points 1′ through 12′ and 1′ again.

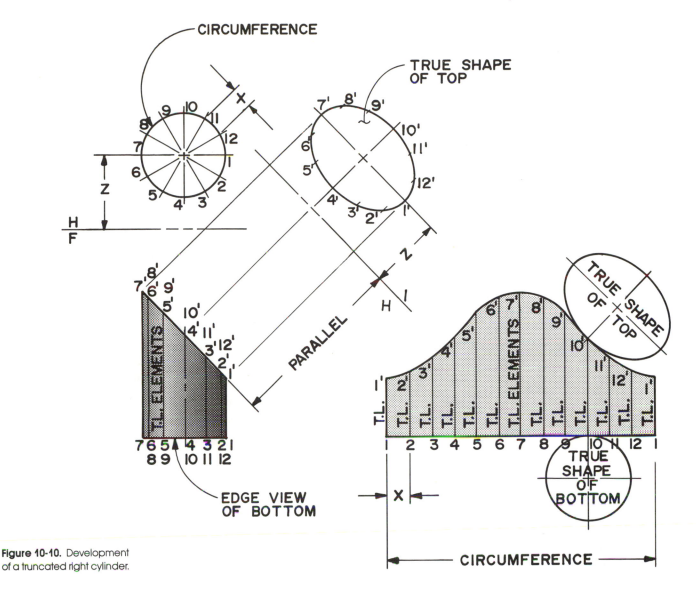

Figure 10-10. Development of a truncated right cylinder.

5. Connecting these points with a smooth curved line completes the lateral surface development.

6. The true shape of the bottom appears in the horizontal view and may be attached at any convenient tangent point. And finally, the true shape of the top appears in view 1, and may be attached at any convenient tangent point on the upper part of the development.

10.3 RADIAL-LINE DEVELOPMENTS

All pyramids and cones are made with radial-line developments. The lateral edges of a pyramid and the elements of a cone appear on their respective developments as straight lines that radiate from a point that corresponds to the vertex of the pyramid or cone. When developing truncated pyramids or cones, it is simpler and more accurate to develop the whole pyramid or cone, and then deduct the vertex portion from the whole.

10.3.1 Development of a Truncated Right Pyramid

When developing a pyramid, the true shape of each face of the pyramid must be determined. Each face of a right pyramid is a triangle with a common vertex, **V**, which is the vertex of the pyramid. The resulting development will be a series of triangles arranged to give the desired form when folded. In a right pyramid, all lateral edges, from the vertex to the base, are the same length. In the example shown in Figure 10-11, none of the lateral edges are true length in the given views. Using the axis of the pyramid as an axis of revolution, edge V-1 can be revolved to appear true length in the front view, V_{F1R}. Line V_{F1R} is the true length of all four lateral edges.

The procedure for developing this pyramid is as follows:

1. Select a starting point, **V**, and construct an arc having the true length of lateral edge, V_{F1R}, as its radius and **V** as its center.
2. On this arc, starting with point 1, on the shortest lateral edge, measure the true length distances around the base, 1 to 2, 2 to 3, 3 to 4, and 4 to 1.
3. Connect point **V** to each of points 1, 2, 3, 4, and 1 respectively.
4. In the front view, measure the distances from V_F to 1′R, 2′R, 3′R, and 4′R, and transfer these distances to their respective lines V-1, V-2, V-3, V-4, and V-1. Points 1′, 2′, 3′, 4′ and 1′ have been located.
5. Connect points 1′ through 4′ and 1′ with straight lines to complete the surface development.

10.3.2 Development of a Truncated Oblique Pyramid

The method for developing an oblique pyramid is essentially the same as that for a right pyramid, except that the lateral edges of the oblique pyramid are different lengths. Because of this difference, the true lengths of each edge must be found. Again, revolution is the most efficient method. In Figure 10-12, two views of a truncated oblique pyramid and its development are shown.

The following procedure is used to develop the surface of this pyramid:

1. Using the vertex, v_H, as an axis of rotation, find the true length of each lateral edge. The true lengths appear in the front view as V_{F1R}, V_{F2R}, V_{F3R}, and V_{F4R}.

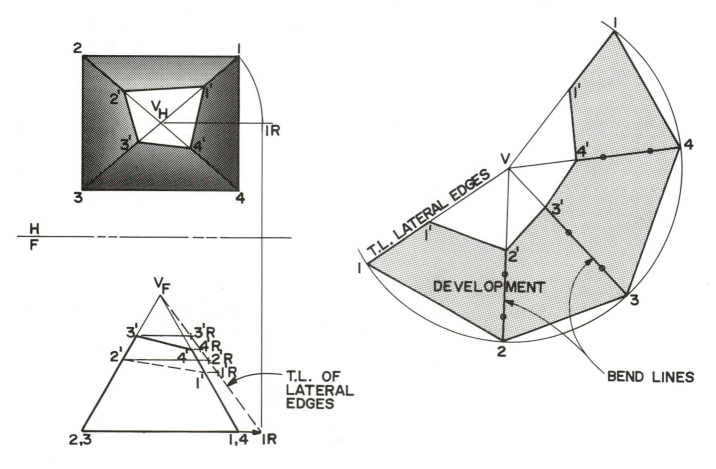

Figure 10-11. Development of a truncated right pyramid.

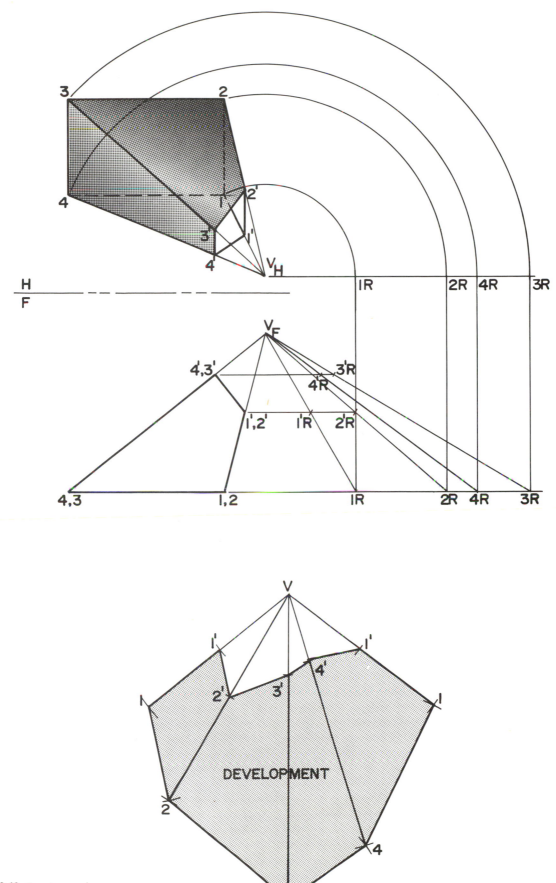

Figure 10-12. Development of a truncated oblique pyramid.

2. Select any point **V** as the vertex for the development. Draw an arc equal to V_{F1R}, the shortest lateral edge. Point 1 is on the arc and its location on the arc is arbitrary.

3. From point 1 draw an arc equal to side 1-2 of the bottom, which is seen true shape in the top view. Then draw an arc equal to V_{F2R} from **V** to establish point 2.

4. From point 2 draw an arc equal to side 2-3 of the bottom. From **V** draw an arc equal to V_{F3R}. Where this arc intersects arc 2-3, you can locate point 3 on the development.

5. From point 3 draw an arc equal to side 3-4 of the bottom. From **V** draw an arc equal to V_{F4R} until it intersects the arc 3-4, thus locating point 4.

6. Now, from point 4 draw an arc equal to side 4-1 of the bottom. From **V** draw an arc equal to edge V_{F1R} until it intersects the arc 4-1. This locates the final point 1.

7. Connect points 1, 2, 3, 4, and 1 with straight lines.

8. From V_F in the front view transfer the true length distances of $V_{F1'R}$, $V_{F2'R}$, $V_{F3'R}$, and $V_{F4'R}$ to their respective lines V-1, V-2, V-3, V-4, and V-1 of the development. Connect points 1′, 2′, 3′, 4′ and 1′ with straight lines. This completes the surface development, exclusive of the top and bottom surfaces.

9. If you need to develop the top and bottom surfaces, the true shapes of each could be attached to the existing development.

10.3.3 Development of a Right Truncated Cone

In Figure 10-13 the orthographic drawing of a right truncated cone and its development are shown. All elements of the cone, vertex to base, are equal in length. Consequently

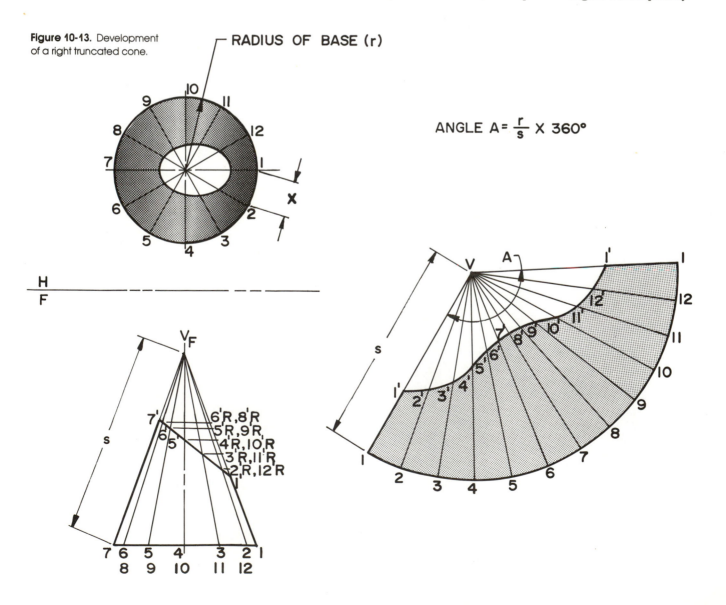

Figure 10-13. Development of a right truncated cone.

RADIUS OF BASE (r)

ANGLE $A = \dfrac{r}{s} \times 360°$

the radial lines of the development will each be a constant length equal to the slant height of the cone. The development will form a circular arc, with **V** as its center, with the length of the arc equal to the circumference of the base of the cone.

The procedure for drawing the lateral surface development of this cone is as follows:

1. The circumference of the base, as seen in the horizontal view, is divided into 12 equal parts, establishing 12 elements on the cone surface. Number the points 1 through 12. Transfer the elements to the front view and label them.

2. Elements V_{F7} and V_{F1} are true length in the front view, and are equal to the slant height of the cone. Now, using **V** as the center and the slant height as the radius, select a convenient starting point and draw an arc of indefinite length.

3. You may determine the length and divisions of this development in two ways: (a) by stepping off the chord distance, **X**, along the arc for the 12 divisions, or (b) by calculating the angle **A** according to the formula given in Figure 10-13. Radial lines should be drawn from **V** to each division point on the arc to represent the elements of the cone.

4. To develop the truncated portion of the cone, you must mark off on each element the true length of the segment cut from the element. Remember, V_{F1} is drawn true length and can be marked off on element V-1. Element V_{F2} is cut at 2′, with its true length seen at $V_{F2'R}$. Similarly, element V_{F3} is cut as 3′, with its true length seen at $V_{F3'R}$. The remaining true length segments are found in the same way, and all are transferred consecutively to their respective elements on the development.

5. Points 1′, 2′, 3′, etc., of the development are connected with a smooth curve to complete the development.

10.3.4 Development of an Oblique Cone

Unlike the right cone, the elements of the surface of the oblique cone are **not** all of equal length. Therefore, you must find the true length of each element before you can begin the development. Figure 10-14 illustrates two views of a truncated oblique cone and its development. The development of such a cone assumes that the surface between any two consecutive elements forms a narrow plane triangle that is a very close approximation to the true shape of this surface. It is very much like an oblique pyramid with 12 sides.

The procedure for constructing the development is as follows:

1. Divide the circumference of the base, as seen in the horizontal view, into 12 equal parts, and number them beginning with the shortest element. Extend all elements to the vertex in the horizontal view. Project the points on the base down to the front view, and extend all elements to V_F.

2. Using the revolution method, determine the true lengths of all elements V-1 through V-12. It is usually more convenient to rotate the elements away from the views, and project the true lengths to the front view, forming a **true-length diagram**, as shown.

3. In addition, by revolution, find the location of points 1′ through 12′ on their respective elements in the true-length diagram. The true lengths from V_F to the prime-numbered points are numbered 1′R, 2′R, 3′R, etc.

4. Begin with the shortest element, V_{F1R}, and a convenient location of **V**. Draw an arc from **V** with a radius equal to V_{F1R} and mark point 1 on the arc.

5. Using point 1 as a center, draw an arc with a radius equal to chordal distance **X**. Then using true-length element V_{F2R}, draw an arc from **V**. Where this arc intersects the chordal-distance arc from point 1, locate point 2. Repeat this process until all points are located on the development. Connect the numbered points with a smooth curve.

6. From the true-length diagram take the true-length distance from V_F to the prime-numbered points on the upper portion of the cone ($V_{F1'R}$, $V_{F2'R}$, etc.) and transfer these distances to the respective elements on the development. The development will be completed when these prime-numbered points are connected with a smooth curve.

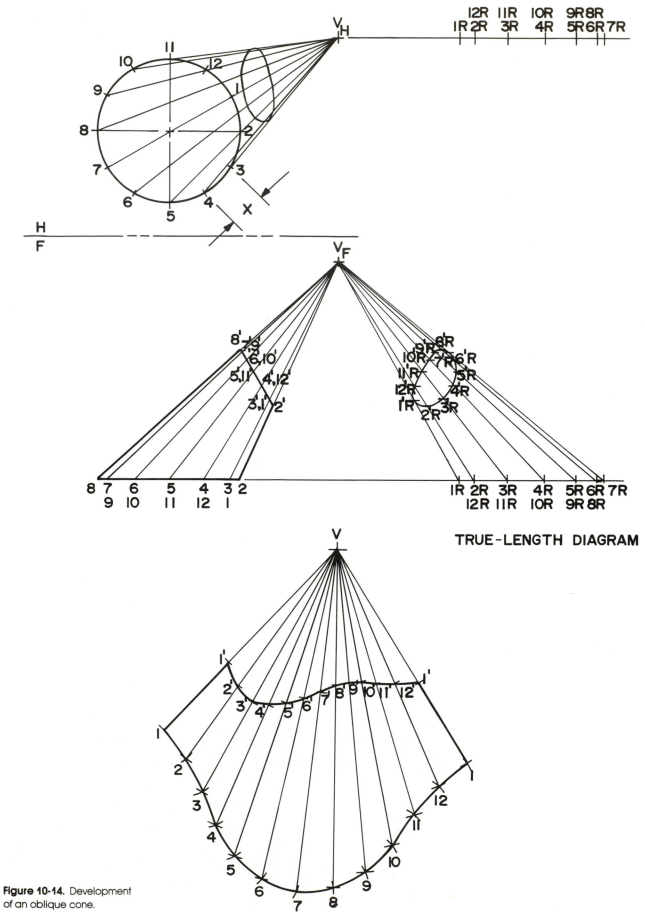

TRUE-LENGTH DIAGRAM

Figure 10-14. Development of an oblique cone.

10.4 TRIANGULATION DEVELOPMENT

Often there are surfaces that cannot be developed by the parallel- and radial-line methods. Developments for many of these surfaces may be constructed using the triangulation method. When using triangulation, the plane surfaces are divided into a series of connected triangular areas. Each triangle in the development can be laid out once the true length of each side has been determined.

10.4.1 Development of a Nonpyramidal Sheet Metal Connector

In Figure 10-15 a sheet metal connector, joining two offset rectangular ducts, is shown orthographically, along with its development. Note that it is not a pyramid, because its edges have no common vertex.

The process for constructing the development of this connector is as follows:

1. In the top view, the diagonal 1-4′ is drawn to divide surface 1-1′-4′-4 into two triangles. Similarly, diagonals 2-1′, 3-2′, and 4-3′ are drawn to divide the other three sides of the connector into triangles. The four diagonals are also drawn in the front view. The connector surfaces now consist of eight triangles.
2. The true length of each edge and its diagonal must be found. Although each could be revolved individually about a different axis to find its true length, it is more efficient and less cluttered to assume a common axis. A true-length diagram has been constructed in which the true length of each edge has been revolved about vertical axis **WX**. The horizontal span, **A**, for edge 1-1′ is marked off as shown on the true-length diagram, and the vertical rise, **E**, is transferred from the front view to provide the true length of 1-1′. This process is repeated for edges 2-2′, 3-3′, and 4-4′. Study this figure carefully.
3. In a like manner, the true lengths of the diagonals are revolved about vertical axis **YZ**. The top-view lengths (horizontal span) are marked off as shown, and the vertical rise, **E**, is transferred from the front view to provide the true lengths of the diagonals.
4. Beginning the development with the shortest edge, the true length of edge 1-1′ is drawn. From point 1, an arc equal to the true length of 1-2 (seen in top view) is drawn. An arc equal to diagonal 2-1′ is drawn from point 1′. The intersection of these two arcs establishes point 2. From point 2 an arc equal to the true length of edge 2-2′ is drawn, and from point 1′ an arc equal to the true length of 1′-2′ is drawn. The intersection of these arcs locates point 2′. The first face of the connector is complete.
5. The remaining triangles of the development are constructed in the same way and in consecutive order, with the needed distances being taken from the top view and the true-length diagrams. The development ends with edge 1-1′, since it began with this edge. Be careful to place the diagonals in the development in the same position as they are in the orthographic views.

10.4.2 Development of a Transition Piece

A sheet metal connector that joins two other pipes, or openings, that are different sizes and shapes is called a transition piece. Figure 10-16 illustrates a transition piece that connects a circular to a rectangular duct. The connecting surface is composed of four triangular planes, **W**, **X**, **Y**, and **Z**, and four partial oblique cones, **R**, **S**, **T**, and **U**.

The procedure for constructing this development is as follows:

1. Divide the circular end into a convenient number of equal parts. Sixteen divisions are used in this example. Cone elements are drawn from each division mark in one-quarter of the circle to the nearest vertex on the rectangle. For example, elements are drawn from 1, 2, 3, 4, and 5 to the vertex at **B**. The true lengths of these elements must be found. Because the surface is symmetrical about centerline A-A′, only one half of the elements are actually rotated to develop the true-length diagrams.
2. In order to find the true lengths of the elements from **B**, they are rotated about **B**. The true lengths of elements B-1, B-2, B-3, B-4, and B-5 are seen in the true-length

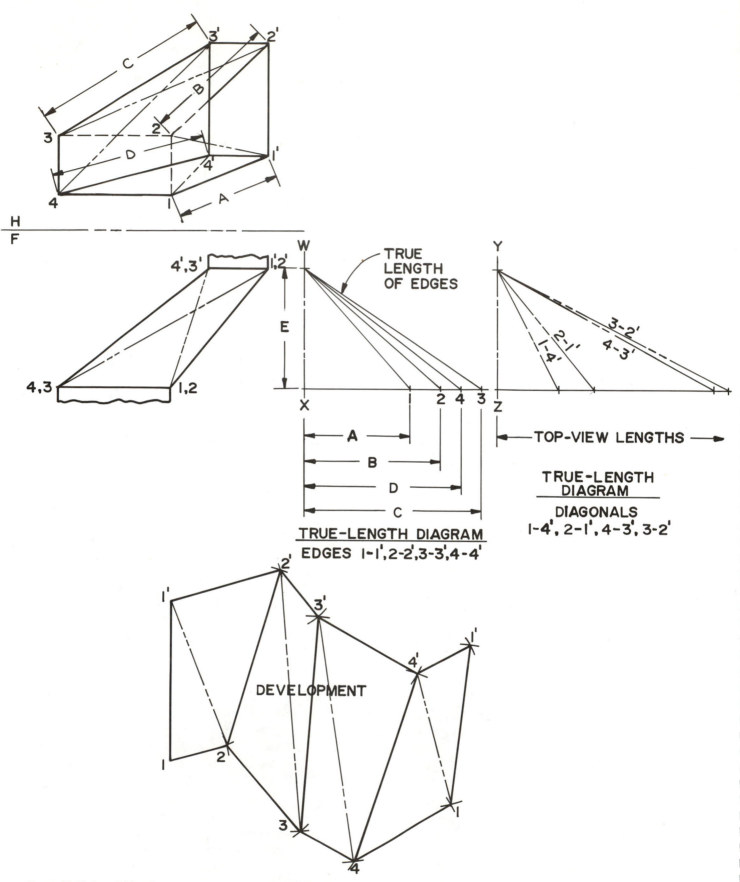

Figure 10-15. Development of a nonpyramidal sheet-metal connector.

diagram next to the front view. Because of symmetry, these lengths will be the same for cone elements E-1, E-16, E-15, E-14, and E-13, respectively.

3. The elements radiating from **C** are rotated about **C** in the horizontal view. The true lengths of C-5, C-6, C-7, C-8, and C-9 are seen in the true-length diagram. These true lengths are equal to the cone elements D-13, D-12, D-11, D-10, and D-9 respectively.

4. The development begins along the centerline at A-1. The true length of A-1 appears in the front view. Draw a line equal to the length of A-1. From point **A**, draw an arc equal to the true length of A-B, and from point 1, draw an arc equal to the true length of element B-1. The intersection of these arcs locates corner **B**.

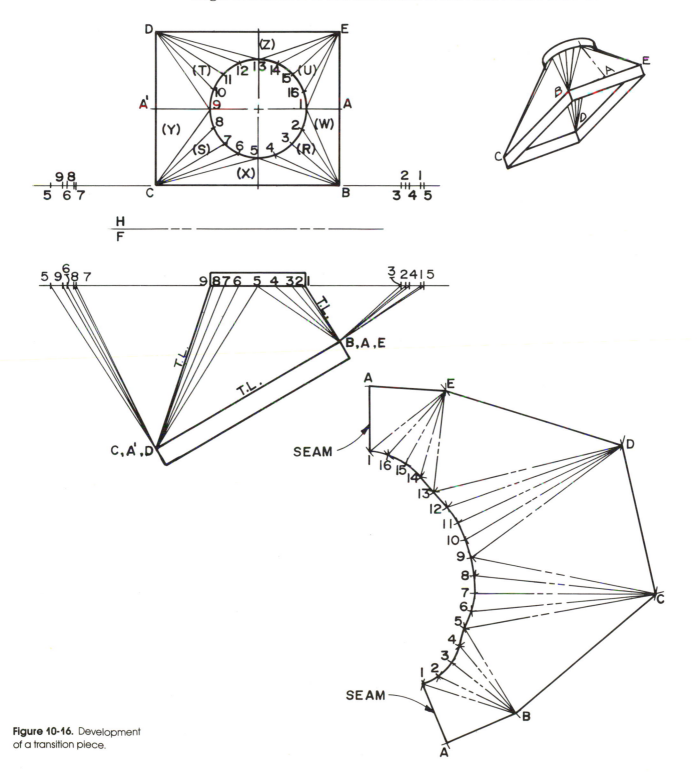

Figure 10-16. Development of a transition piece.

5. An arc equal to the chordal distance across one division on the circular opening is drawn from point 1. From **B**, an arc equal to element B-1 is drawn. This locates point 2. This same process is repeated to locate elements B-3, B-4, and B-5.

6. From point **B**, an arc equal to the true length of side B-C of the rectangle is drawn. Note that the true length of B-C appears in the front view. From point 5, an arc equal to the true length of element C-5 is drawn. The intersection of these arcs establishes point **C**. With point **C** located, the elements radiating from **C** can be drawn.

7. This process continues until you have completed the development by returning to the seam at A-1.

10.4.3 Approximate Development of a Warped Transition Piece

In Figure 10-17 a transition piece connecting two nonparallel circular openings is shown. This is a warped surface. A development that is a very close approximation of the actual size of this piece can be constructed using the triangulation method.

The procedure for constructing this development is as follows:

1. To develop this warped surface, elements equally spaced on each opening are used. Equal spaces on the large horizontal opening can be constructed in the true-shape top view. On the inclined opening, the equal spaces can be found by drawing a semicircle on the front view and dividing it into six equal parts. This serves as a substitute for an auxiliary view. The like-numbered elements on each opening should be connected, 1-1′, 2-2′, etc.

2. Diagonals, 1-2′, 2-3′, 3-4′, etc., are then drawn in each view between adjacent elements. Two true-length diagrams, one for elements, and one for diagonals, are drawn. These are similar to the ones drawn in Figure 10-15. Each top-view length is laid off from the axis of rotation directly in line with each element or diagonal. For diagonal 1-2′, the top-view length, **Y**, is laid off in line with 2′ to the right of the axis of rotation. Notice that the true-length elements are shown to the left of the axis, and the true-length diagonals to the right of the axis, for clarity.

3. Only a half development is shown in the illustration because the connector is symmetrical. The construction of the development begins with the shortest element, 1-1′, and proceeds one triangle after another.

4. Lay out element 1-1′ in a convenient location. From point 1′, draw an arc equal to the chordal distance, 1′-2′, from the semicircular view attached to the smaller opening. From point 1, draw an arc equal to the true-length diagonal 1-2′. The intersection of these arcs locates point 2′.

5. From point 1, draw an arc equal to the chordal distance between equal divisions on the large opening, and from point 2′, draw an arc equal to element 2-2′. The intersection of these arcs establishes point 2.

6. This process is continued until element 7-7′ is drawn. Connect points 1, 2, 3, 4, etc., with a smooth curve, and do the same with points 1′, 2′, 3′, 4′, etc. to complete the half development.

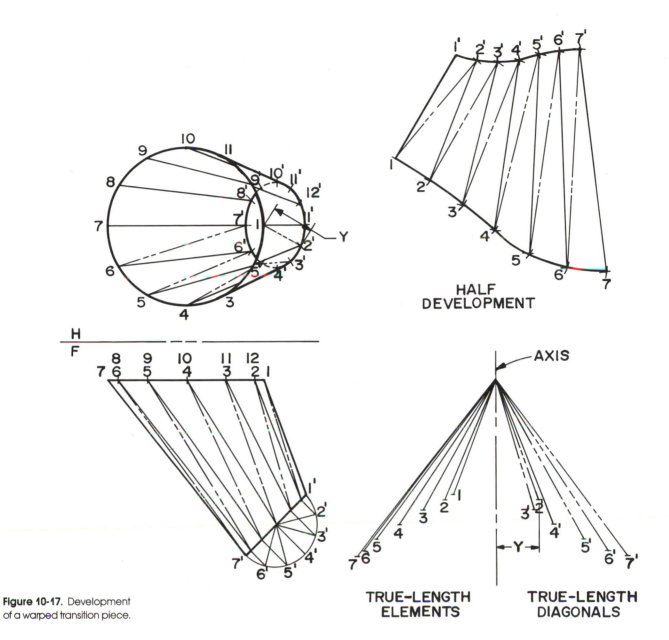

Figure 10-17. Development of a warped transition piece.

HALF
DEVELOPMENT

AXIS

TRUE-LENGTH
ELEMENTS

TRUE-LENGTH
DIAGONALS

CHAPTER PROBLEMS

Problem 1: Draw the given orthographic views and develop the lateral surfaces of the right prism shown below. Begin your development with the shortest edge. Use a B-size (11"×17") drawing sheet. Scale: Full

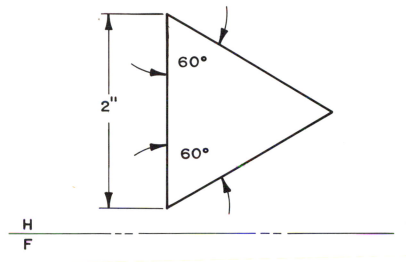

60°

2"

60°

H
F

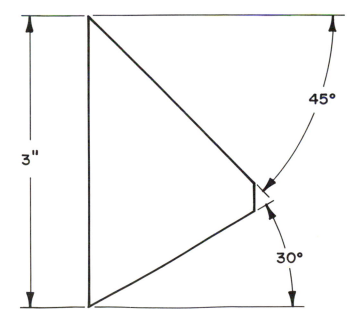

3"

45°

30°

Problem 2: A package chute is to extend from the rear wall through the floor and is to make a 30° angle with the floor. On a C-size drawing sheet, show the given views. Complete the front view. Develop the lateral surfaces of the chute beginning with the shortest edge. Scale: 1" = 3'

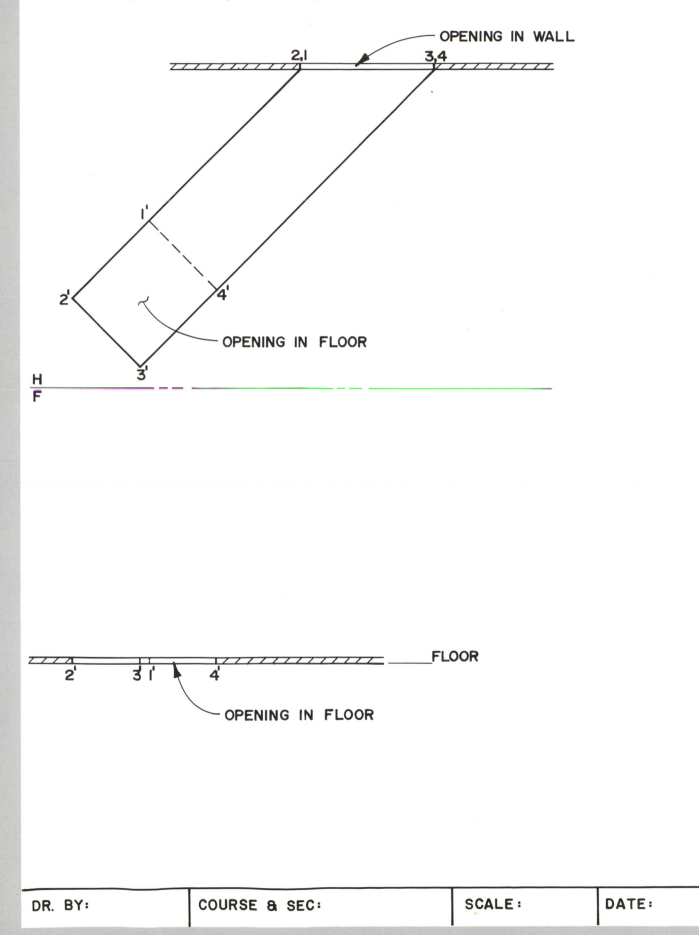

OPENING IN WALL

2,1   3,4

1'

2'

4'

OPENING IN FLOOR

3'

H
F

FLOOR

2'   3 1'   4'

OPENING IN FLOOR

DR. BY:	COURSE & SEC:	SCALE: 1" = 3'	DATE:

Problem 3: Develop the lateral surface of the right cylinder shown below. Begin with the shortest element on the given stretch-out line. Scale: Full

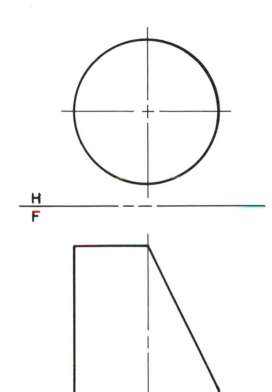

H
F

STRETCH-_____ __ _____
OUT LINE

DR. BY:	COURSE & SEC:	SCALE:	DATE:

Problem 4: An 8-inch diameter connector is shown below. On a B-size drawing sheet, develop the lateral surface of this connector. Scale: 1" = 8"

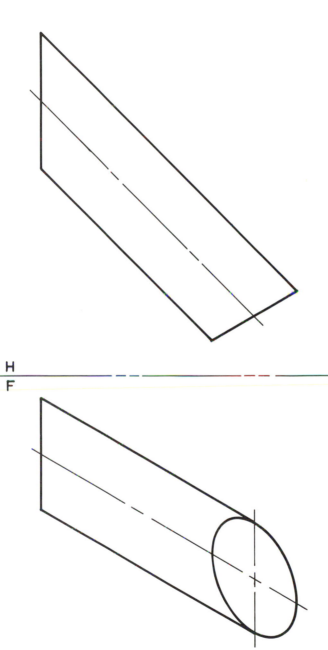

H

F

Problem 5: Draw a development of the pyramidal piece shown below. Scale: 1" = 10"

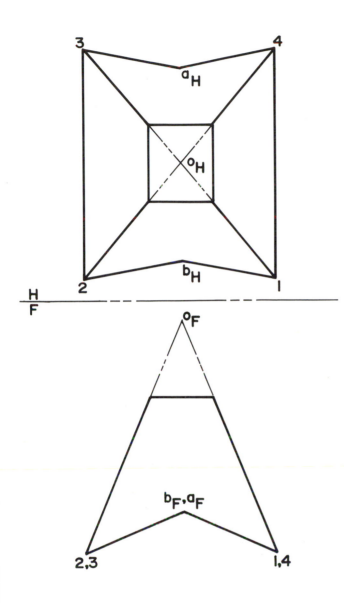

Problem 6: On a B-size drawing sheet, provide a development of the pyramidal portion of the sand hopper shown below. Show your true-length diagram on this sheet. Scale: ½" = 1'-0"

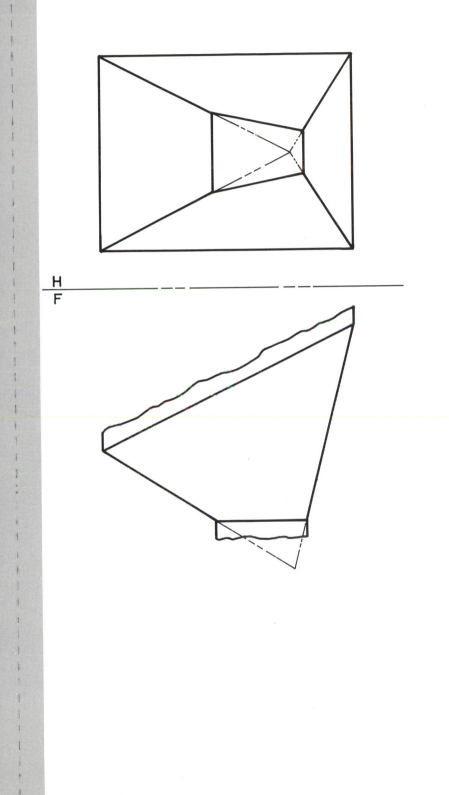

H
F

| DR. BY: | COURSE & SEC: | SCALE: | DATE: |

Problem 7: Complete the top view and develop the lateral
surface of the right circular cone shown below. Scale: 1" = 2"

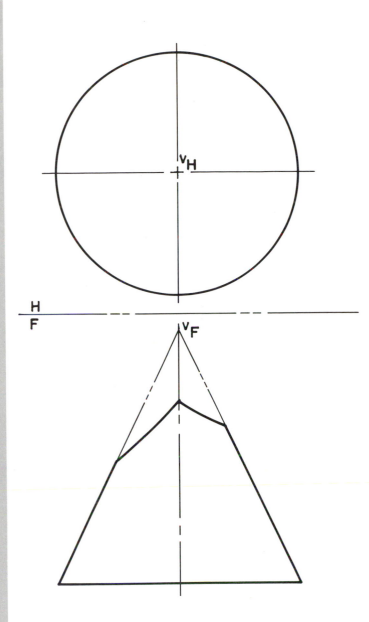

H
F

V_H

V_F

DR. BY: COURSE & SEC: SCALE: DATE:

Problem 8: Find the development of the oblique, conical connection shown below. Begin with the shortest element.
Scale: 1" = 4"

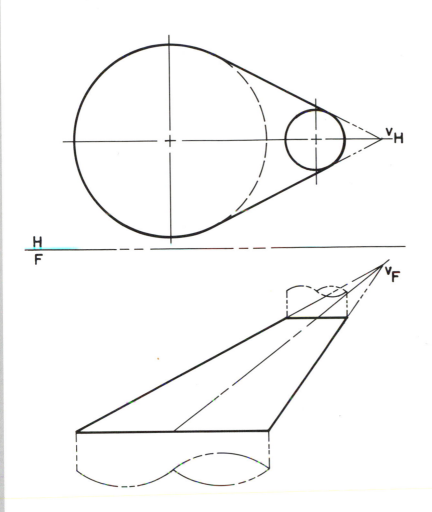

H
F

V_H

V_F

Problem 9: Develop the lateral surfaces of this nonpyramidal transition piece. Show the necessary true-length diagrams on this worksheet. Construct your development on a B-size drawing sheet. Scale: 1" = 1'-0"

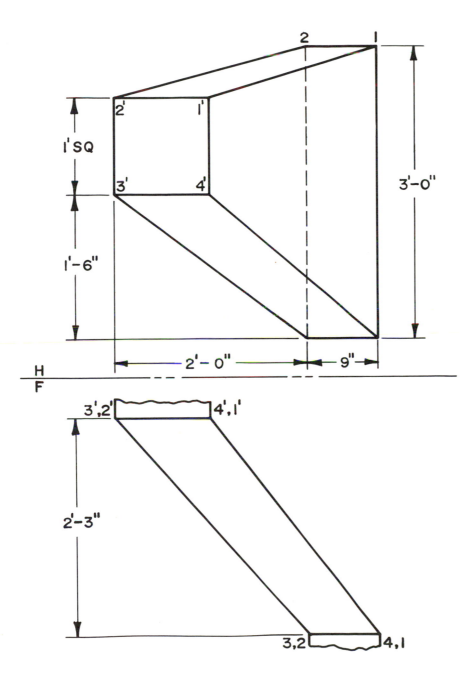

1' SQ

3'-0"

1'-6"

2'-0"

9"

H
F

3',2'   4',1'

2'-3"

3,2   4,1

Problem 10: Develop the lateral surfaces of the transition piece connecting the rectangular fireplace box to the cylindrical chimney flue. Show the given views, true-length diagrams, and your development on a C-size drawing sheet. Use the triangulation method and begin with the shortest element. Scale: ½" = 1'-0"

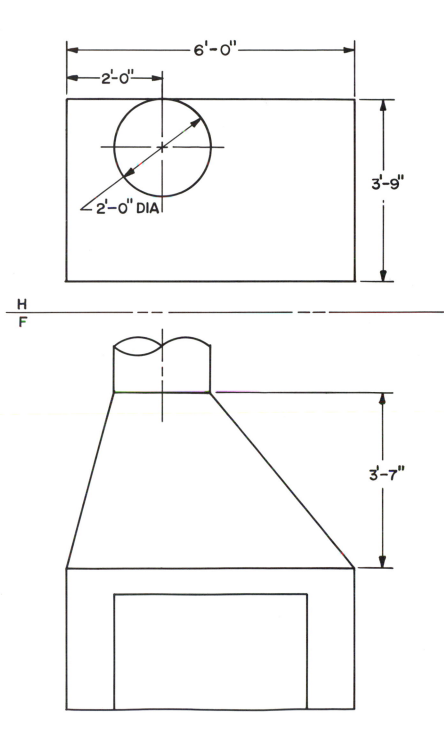

Problem 11: Develop the lateral surfaces of the sheet-metal-duct transition piece. Show the given views, the true-length diagram, and the development on a B-size drawing sheet.

Use the triangulation method and begin with the shortest element. Scale: 1" = 1'-0"

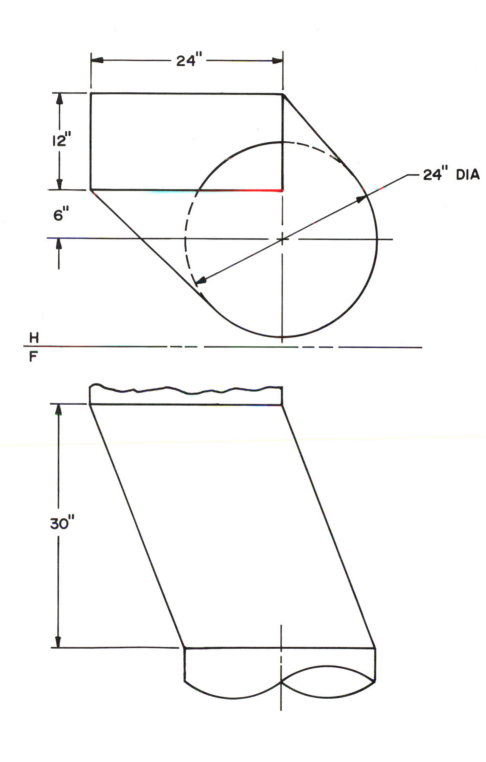

DR. BY:	COURSE & SEC:	SCALE:	DATE:

Problem 12: A transition piece connecting circular openings in different planes is needed. Show the true-length diagrams on this worksheet. Use triangulation to develop the lateral surfaces of this connector on a B-size drawing sheet. Scale: 1" = 2"

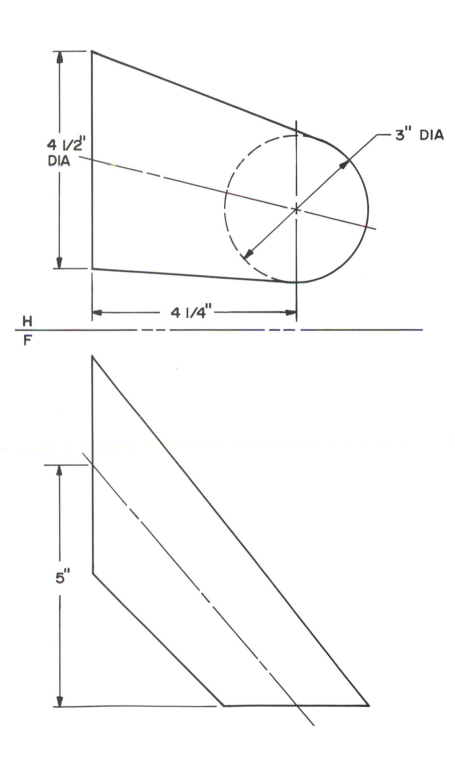

| DR. BY: | COURSE & SEC: | SCALE : | DATE : |

CHAPTER TEN

TEST

1. What is meant by the development of a surface?

2. Explain how developments are used in actual practice.

3. Developments are divided into three general groups. List them.

4. Contrast parallel-line and radial-line developments.

5. Without using a sketch, briefly describe the following:
   a. Stretch-out line:

   b. Truncated prism:

   c. Element:

   d. Bend line:

   e. Transition piece:

6. How must all the lines to be used in a development appear before they can be transferred to the development?

7. Why are developments generally joined along the shortest edge, or element?

8a. When developing an oblique prism, what orthographic views would you need?

8b. Give a concise explanation of the procedure you would use for constructing the development of an oblique prism.

9. The development of a right cylinder has what shape?

10. How is the development of a pyramid similar to that of a cone?

11. What is a true-length diagram with regard to developments?

12. Describe the steps you would take to develop a truncated right prism. (Hint: A list of steps to follow in all cases of truncated right prisms)

13. Explain the triangulation method of development.

14. Explain how you would develop a square-to-round transition piece. (You may use a sketch to accompany your explanation.)

CHAPTER
ELEVEN

Vector Geometry

11.1 INTRODUCTION

An understanding of vector geometry is important, because vectors are a vital part of many engineering sciences, such as mechanics and kinematics. The mathematical manipulation of vectors is dealt with in a branch of mathematics called **vector analysis**. This chapter deals with the graphic manipulation of vectors which is a fast and sufficiently accurate method for many purposes. Vectors can be manipulated in some very complex ways, but the discussion here is limited to vector addition and its practical applications.

A vector quantity is one that requires both magnitude and direction for its complete description. Force, velocity, acceleration, displacement, and momentum are examples of vector quantities because they require size and direction to be complete. For example, if a car is said to be traveling at 55 miles per hour, it is a statement of speed, but if it is said to be traveling at 55 miles per hour in a due south direction, it is a statement of velocity, which is a vector quantity.

11.2 VECTOR REPRESENTATION

A vector quantity can be represented by a straight line whose length is equal to the magnitude of the quantity to some chosen scale. The direction is along the line and its sense is identified by an arrowhead. In Figure 11-1, you can see two vectors, **A** and **B**, shown in a **space diagram** in which the vectors are drawn in the correct direction and sense, but **not to scale**.

When two or more vectors act on an object, it is called a **vector**, or **force system**. A vector system may be **coplanar**, meaning that all the vectors lie in the same plane. Figure 11-2 shows four coplanar vectors, pictorially and orthographically. In addition to being coplanar, the system of vectors in Figure 11-2 all act through a common point, **O**. Such a system is **concurrent**. Figure 11-3 illustrates pictorially and orthographically, a concurrent, **noncoplanar** system of vectors. The vectors act in different planes, but they all meet at a single point, **O**.

SPACE DIAGRAM

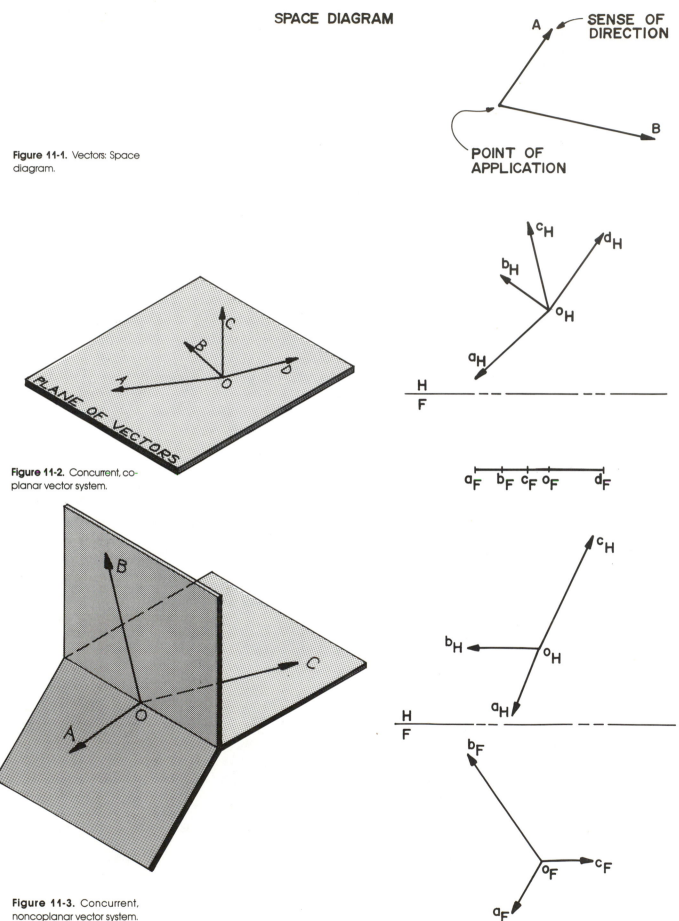

Figure 11-1. Vectors: Space diagram.

Figure 11-2. Concurrent, coplanar vector system.

Figure 11-3. Concurrent, noncoplanar vector system.

Forces may be **nonconcurrent**, that is, their lines of action do not meet at one point. Such systems can be coplanar or noncoplanar. Figure 11-4 illustrates a system of nonconcurrent, noncoplanar, parallel vectors. Figure 11-5 illustrates a nonconcurrent, noncoplanar, unparallel vector system.

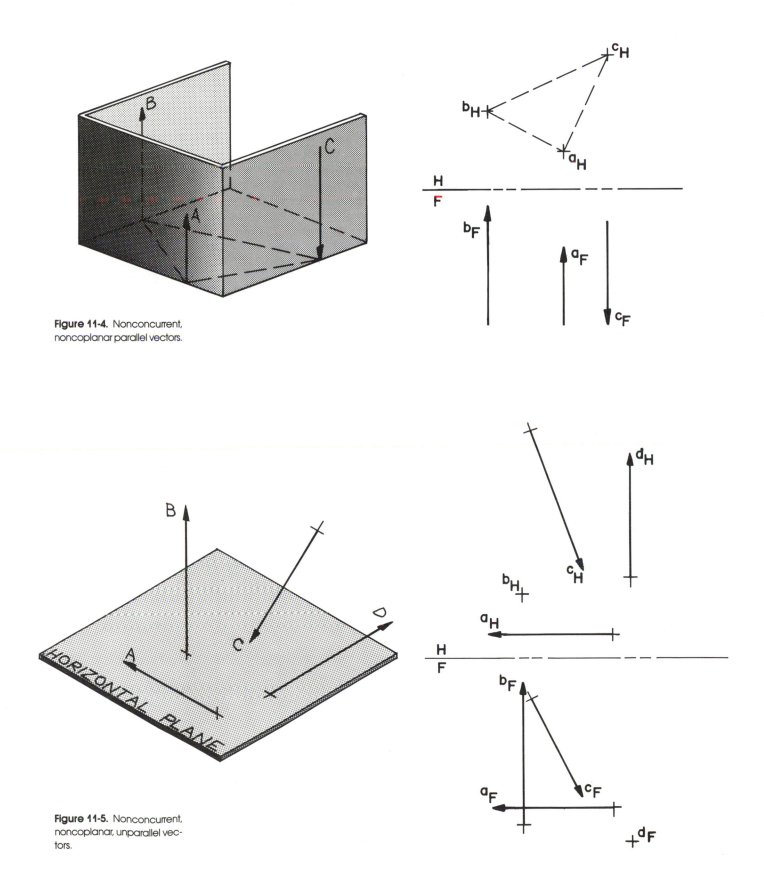

Figure 11-4. Nonconcurrent, noncoplanar parallel vectors.

Figure 11-5. Nonconcurrent, noncoplanar, unparallel vectors.

11.3 ADDITION OF CONCURRENT, COPLANAR VECTORS

Vector quantities must often be combined, or added, to determine the net effect of a vector system. The graphic addition of concurrent, coplanar vectors is simply a matter of placing vectors together, tails to heads, constructing the correct magnitude and maintaining the direction indicated in the space diagram. In the **vector diagram**, the first vector is drawn and the tail of the second vector is connected to the head of the first vector. The closing vector, or the sum of the vectors, is drawn from the tail of the first vector to the head of the second vector, with its direction indicated, as seen in Figure 11-6. This sum is called a **resultant**, and has the same effect as the total of the individual vectors that have been added.

If the direction of the resultant is reversed, it is called the **equilibrant**. The equilibrant has the effect of opposing the added vectors to place the system in exact balance, as seen in Figure 11-7. An equilibrant is the same as a resultant of zero.

Figure 11-8 shows the addition of the same vectors, **A** and **B**, by the **parallelogram method**. You can see that the same resultant is obtained as with the **vector polygon** (tails to heads) method just described. The parallelogram method requires more construction in that the sides of the parallelogram are drawn parallel to the given vectors and then a diagonal equal to the resultant is drawn. When more than two vectors must be added, the vector polygon method is preferred.

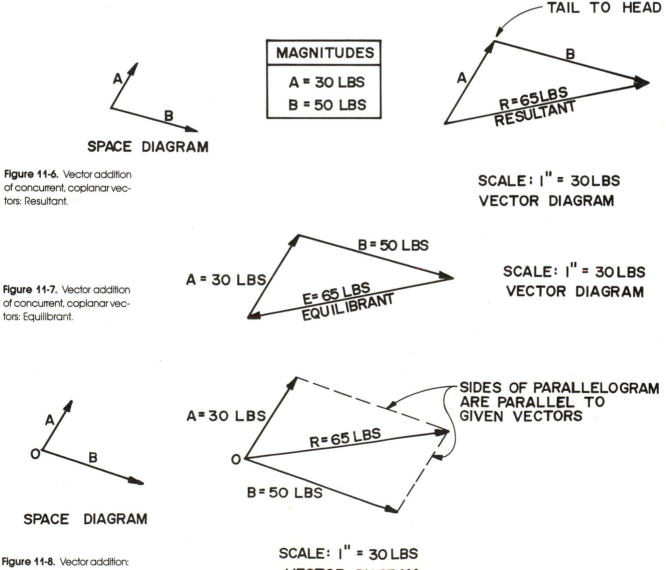

Figure 11-6. Vector addition of concurrent, coplanar vectors: Resultant.

Figure 11-7. Vector addition of concurrent, coplanar vectors: Equilibrant.

Figure 11-8. Vector addition: Parallelogram method.

Three or more vectors can be added using the same methods. The given vectors (see Figure 11-9[a]) are drawn to scale and parallel to the given direction sequentially in a tail-to-head manner, until a chain of vectors is formed. The closing side of the polygon formed by the vectors, drawn from the tail of the first vector to the head of the last vector, with the direction indicated, as seen in Figure 11-9(b), is the resultant. If the equilibrant is required, then the closing vector will go in the direction of the head of the last vector to the tail the first vector. In this example, the equilibrant would be drawn in the opposite direction from the resultant seen in Figure 11-9(b).

Often coplanar and concurrent vectors act in an oblique plane. When such a case occurs, the true magnitude of the vectors is not seen in the primary views. To find the resultant vector, you must treat each view separately as though they were independent vector systems. In Figure 11-10(a) the top and front views of two concurrent, coplanar vectors acting in an oblique plane are shown. In figure 11-10(b) the resultant is found using the vector-polygon method. In the top view vector **X** is drawn parallel to given vector **X** and vector **Y** is added to it. The closing side of the polygon in the top view is the resultant, R_H. The same procedure is followed in the front view. Vector **X** is drawn parallel to the given vector **X**, and vector **Y** is added to it. The closing side of the polygon is the front view of the resultant, R_F. Notice that the alignment of the component arrowheads acts as a check on your accuracy. The last procedure is to rotate resultant R_H until it is parallel to the H/F fold line to obtain the true magnitude of the resultant in the front view.

11.4 RESOLUTION OF CONCURRENT, COPLANAR VECTORS

You have seen that two or more vectors can be added to form a single resultant vector. This procedure can be reversed, and a single vector can be **resolved** into **component** vectors. In the case of coplanar vectors, there can be only two components of unknown magnitude or direction in order for the vectors to be resolved. Most commonly, the direction is known, and the magnitude must be determined.

Figure 11-9. Vector addition: Three concurrent, coplanar vectors.

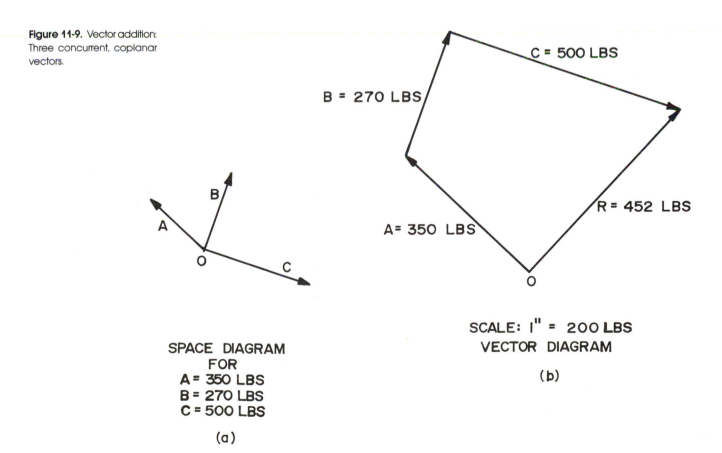

SPACE DIAGRAM
FOR
A = 350 LBS
B = 270 LBS
C = 500 LBS

(a)

C = 500 LBS

B = 270 LBS

R = 452 LBS

A = 350 LBS

SCALE: 1" = 200 LBS
VECTOR DIAGRAM

(b)

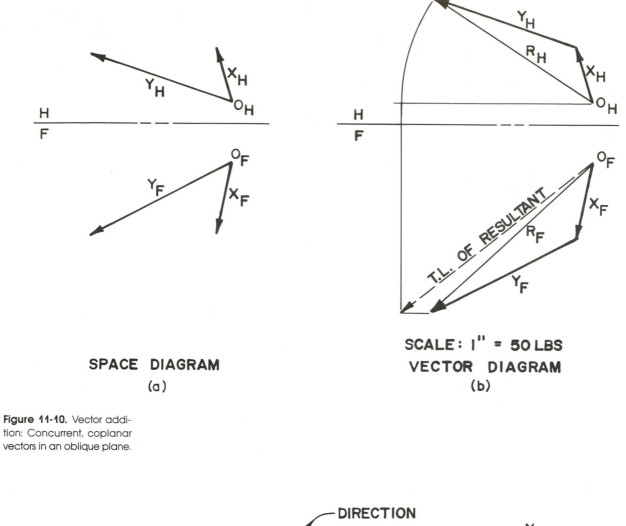

SPACE DIAGRAM

(a)

SCALE: 1" = 50 LBS

VECTOR DIAGRAM

(b)

Figure 11-10. Vector addition: Concurrent, coplanar vectors in an oblique plane.

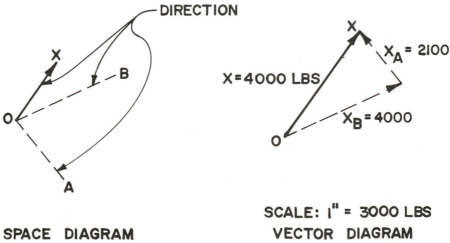

Figure 11-11. Resolution of a single vector into two component vectors.

SPACE DIAGRAM

SCALE: 1" = 3000 LBS

VECTOR DIAGRAM

In Figure 11-11 a given vector **X** must be resolved into two components having the directions **OA** and **OB**. The vector diagram shows vector **X** to its scaled length of 4,000 pounds. From the tail of vector **X** at point **O**, a line is drawn parallel to the given direction **OB**. Through the head of vector **X** another line is drawn parallel to given direction **OA**. The intersection of these lines completes the vector polygon and defines the magnitudes of the component vectors. The actual sense of direction of the components is determined by realizing that they must be tail to head beginning at point **O**.

11.5 ADDITION OF CONCURRENT, NONCOPLANAR VECTORS

The theory of the addition of noncoplanar vectors is the same as for coplanar vectors. As with coplanar vectors in an oblique plane, the principles of descriptive geometry must be used to solve a problem which now occupies three-dimensional space.

In Figure 11-12 the addition of noncoplanar vectors is illustrated. On the left, the top and front views of the given vectors are shown in the space diagram. On the right the addition process is completed in the vector diagram. The space diagram indicates the sense of direction of each vector in each view, and the true magnitudes of the three vectors are listed. In the vector diagram, each vector must be drawn parallel to the corresponding given vector in the space diagram. The scaled length of the vector must be laid off in a view showing the true length of the vector. In the example, vector **A** is first laid out to scale in a direction parallel to **A**$_H$ in the top view, because it appears true length. The front view of vector **A** is found by alignment. In contrast, vector **B** must be scaled to its true length in the front view. The top view of vector **B** is found by alignment. If the vector is oblique, as is vector **C**, its scaled length must be found in a true length view. Vector **C** is drawn in the correct direction in both views, and to some arbitrary end

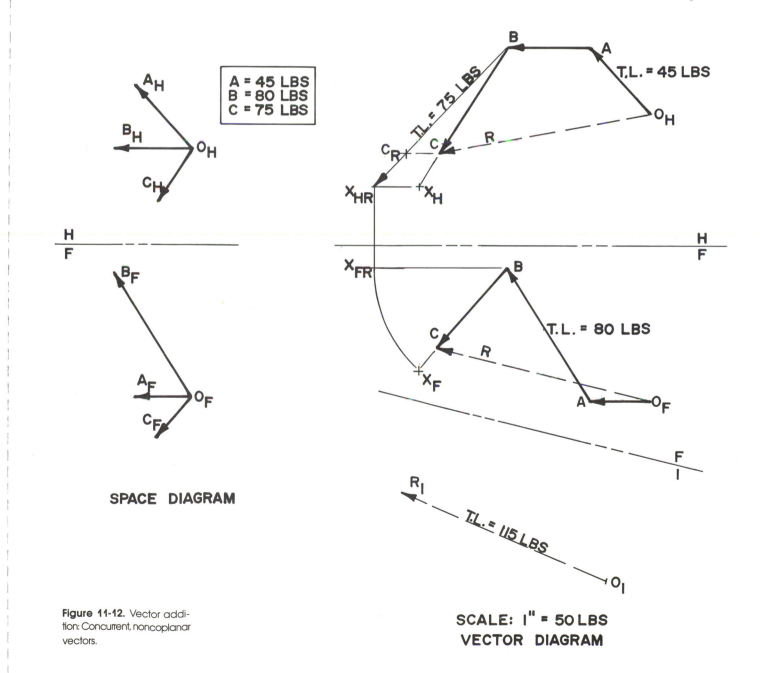

Figure 11-12. Vector addition: Concurrent, noncoplanar vectors.

SPACE DIAGRAM

SCALE: 1" = 50 LBS
VECTOR DIAGRAM

point **X**. Point x_F is rotated to a position parallel to fold line H/F, and a true length line to point x_{HR} is seen in the top view. The correct magnitude of vector **C**, 75 pounds, may be scaled off on this line, and then transferred to the horizontal view of vector **C**. The front view of vector **C** is found by projection. Now the resultant may be found in each view. The auxiliary view method is used to show the actual magnitude of the resultant (see view 1).

11.6 FORCES IN EQUILIBRIUM

As mentioned previously, when a series of forces acting on a particular structure, or body, has a resultant of zero, then this system is said to be in **equilibrium**. Graphically, when the forces are in equilibrium, the vector polygon closes, with vectors in a continous tail to head arrangement, and with a resultant of zero. If a particular body is acted upon by several known and/or unknown forces, and is known to be in equilibrium, you can employ these graphic concepts to determine the unknown forces. It should be noted that the number of unknown quantities must be limited. A vector system may be any number of known forces, but a coplanar vector system can have only two unknown vector quantities, and a noncoplanar system can have only three unknowns in order to be solved. When analyzing equilibrium problems, it is important to isolate the structure, or part, to which the known forces are applied. It is only by doing so that the vector diagram can be constructed.

11.6.1 Coplanar Systems in Equilibrium: Two Unknowns

Here and in subsequent problems, **Bow's Notation** will be used to label the forces. In this type of notation, a letter is given to the space on each side of a vector, and each vector is then identified by the two letters on either side of it, reading in a clockwise direction.

In Figure 11-13 two cables, supporting a 500-pound weight, pass over pulleys and support weights of 300 and 400 pounds. Hence the magnitudes of the forces acting in the cables are known, but the directions of vectors **YZ** and **ZX** for equilibrium must be found. The vector diagram begins with the known vertical force **XY**, scaled to represent 500 pounds. Next, with a radius equal to 300 pounds, an arc is drawn with **Y** as its center, and with a radius equal to 400 pounds, another arc is drawn with **X** as its center.

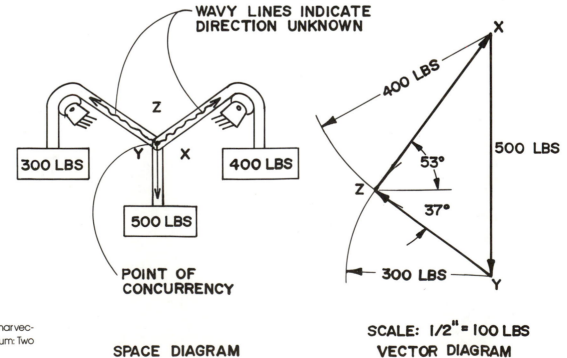

Figure 11-13. Coplanar vector system in equilibrium: Two unknowns.

SPACE DIAGRAM

SCALE: 1/2" = 100 LBS

VECTOR DIAGRAM

These arcs intersect at **Z** to complete the polygon. The directions are indicated by the angles shown. The sense of direction is determined by placing arrowheads in a tail-to-head manner around the diagram. Transfer these arrowhead directions to the space diagram and observe that they are all **pulling away** from the point of concurrency, indicating that the members are all in **tension**.

A simple truss in equilibrium is shown in Figure 11-14. A vertical load of 2,000 pounds is known, and the directions of the forces acting in members **AB** and **BC** are known. It is necessary to find the magnitudes of the forces acting on these members. The vector diagram begins with known force **AB** drawn vertically and scaled to represent 2,000 pounds. Next a line is drawn parallel to the given direction of member **BC**, and another line is drawn parallel to the given direction of member **CA**. The intersection of these two lines at **C** forms the polygon. The magnitudes are found by scaling the vector diagram. The senses of direction of the forces are found by placing arrowheads in a tail-to-head arrangement around the polygon. Let's transfer the arrowheads, the senses of direction, to the space diagram. Notice that the force in **CA** pulls away from the point of concurrency, meaning that the member is in tension. In contrast, the force in member **BC** pushes toward the junction, meaning that the member is in **compression**.

11.6.2 Maxwell Diagrams

We can now carry this process of analyzing forces on a structure one step further. Figure 11-15 shows a roof truss and the known forces occurring on this system.

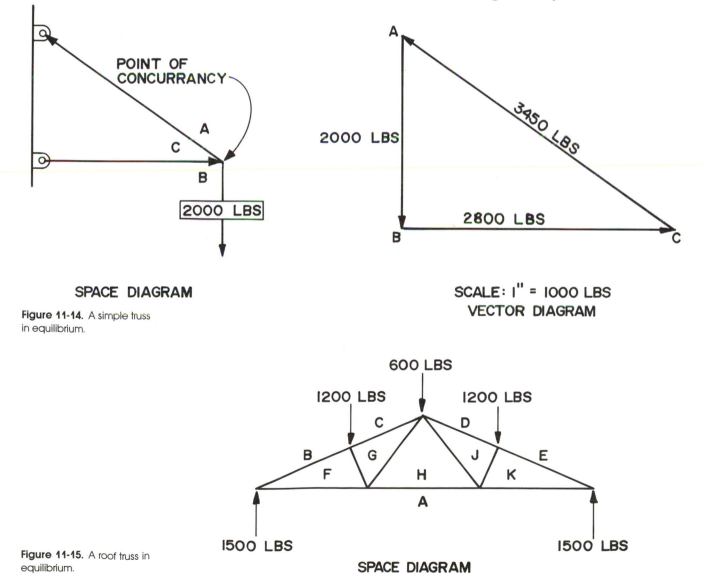

SPACE DIAGRAM

SCALE: 1" = 1000 LBS
VECTOR DIAGRAM

Figure 11-14. A simple truss in equilibrium.

Figure 11-15. A roof truss in equilibrium.

First consider an isolated portion of this truss as seen in Figure 11-16. A vector diagram shows the known force, **AB**, of 1,500 pounds drawn vertically. Next, lines parallel to members **FA** and **BF** are drawn, which intersect at **F**. Knowing the sense of direction for force **AB**, the directional arrowheads are placed on **BF** and **FA** in the tail-to-head manner. As in previous examples, transfer these arrowhead directions to the space diagram. Note that member **BF** pushes toward the point of concurrency, meaning a compressive force, and member **FA** pulls away from this point, indicating a tensile force.

This procedure is identical to that followed in Figure 11-14 and each joint in the truss could be analyzed in this manner. But rather than constructing separate force diagrams for each joint, it is more convenient to combine them into a single diagram, known as a **Maxwell diagram**, as shown in Figure 11-17.

To begin the diagram, select a convenient starting point **A**. The given loads are laid off as follows: **AB** = 1,500 pounds in an upward direction, **BC** = 1,200 pounds in a downward direction, **CD** = 600 pounds in a downward direction, **DE** = 1,200 pounds in a downward direction, and **EA** = 1,500 pounds in an upward direction.

Next, examine the first joint at the left (same as used in Figure 11-16). Point **F** can be located by drawing through **B** a line parallel to member **BF** and another line through point **A**, parallel to member **FA**. The intersection of these lines locates point **F** on the Maxwell diagram. Point **G** is located at the intersection of a line drawn through point **F** parallel to **FG** and a line drawn through point **C** parallel to member **CG**. Point **H** is at the intersection of the line drawn through point **A** parallel to member **AH**, and a line drawn through point **G** parallel to member **GH**. This process is continued to locate points **J** and **K**. Once the force diagram is completed, the magnitudes of the forces can be scaled directly. The determination of the type of force, compressive or tensile, can also be made. For example, examine the peak joint, and begin with the given force **CD**. As you read clockwise about the joint, the sequence is **CD**, **DJ**, **JH**, **HG**, **GC**. Using this sequence in the Maxwell diagram, arrowheads would be placed at **J**, **H**, **G**, and **C**. Applying the direction of the forces to the space diagram, you can see that members **DJ** and **GC** are in compression and members **JH** and **HG** are in tension. The load table of the unknown forces can then be completed with the magnitude and type of force on each member.

11.7 EQUILIBRIUM OF NONCOPLANAR, CONCURRENT FORCES

When a noncoplanar, concurrent system of forces acts on a structure in equilibrium, not more than three forces can have unknown magnitudes for the diagram to be solved. **In any given view of this system, only two unknown magnitudes may be determined.** Two methods of solution are discussed, one in which one unknown force appears as a point, and another in which two unknown forces appear coincident (as an edge).

Figure 11-16. An isolated portion of the roof truss from Figure 11-15.

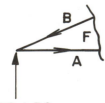

1500 LBS

SPACE DIAGRAM

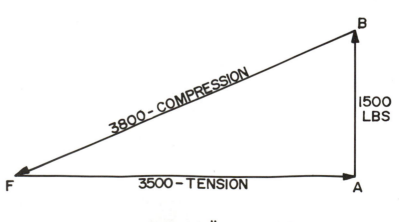

SCALE: 1/2" = 500 LBS
VECTOR DIAGRAM

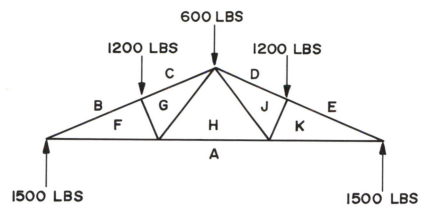

SPACE DIAGRAM

LOAD TABLE
(UNKNOWN FORCES)

MEMBER	MAGNITUDE OF FORCE	TYPE
BF	3800	C
CG	3800	C
DJ	3300	C
EK	3300	C
GF	1100	C
KJ	1100	C
HG	1300	T
JH	1300	T
AK	3500	T
AH	2300	T
AF	3500	T

Figure 11-17. Maxwell diagram.

SCALE: 1/2" = 500 LBS
MAXWELL DIAGRAM

11.7.1 One Unknown Force as a Point

A vertical mast and two guy wires are shown in Figure 11-18. The space diagram indicates that a 1,500-pound load is applied to the system. Bow's notation is used, with spaces labeled clockwise beginning with the known force, **AB**. The spaces are labeled in the horizontal view. Notice that the vertical mast is shown as a wavy line, just to facilitate the notation.

The procedure for the construction of the vector diagram is as follows:

1. Remember that vectors drawn in each view of the vector diagram **must** be parallel to the given directions shown in the space diagram.

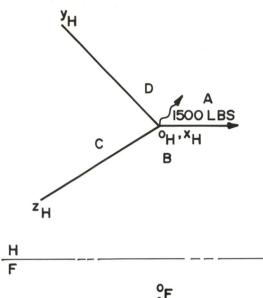

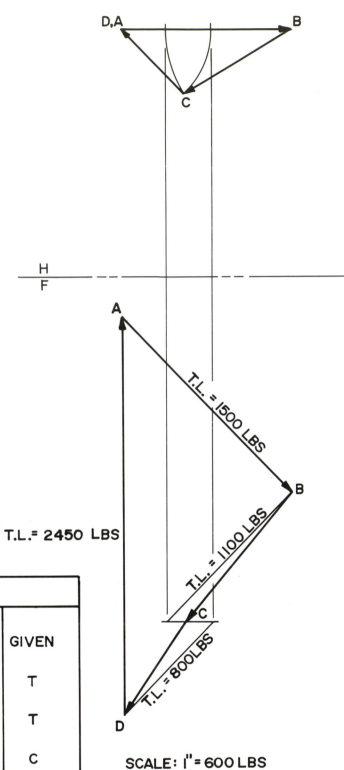

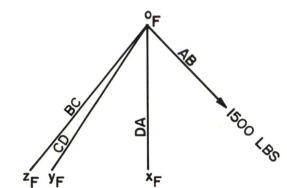

SPACE DIAGRAM

LOAD TABLE		
AB	1500 LBS	GIVEN
BC	1100 LBS	T
CD	800 LBS	T
DA	2450 LBS	C

T.L.= 2450 LBS

SCALE: 1" = 600 LBS
VECTOR DIAGRAM

Figure 11-18. Noncoplanar, concurrent vector system in equilibrium: One unknown force as a point.

2. First, vector **AB** is drawn to scale in the front view where it appears true length. Once the front view is established, the horizontal view can be located by alignment.
3. In the horizontal view, mast **DA** appears as a point, thus only two unknown vectors are seen. Vector **BC** can be drawn through **B** parallel to **BC** of the space diagram, and vector **CD** can be drawn through **D** parallel to **CD** of the space diagram. This closes the vector polygon in the horizontal view.
4. In the front view, vector **BC** can now be drawn with point C located by alignment with the top view. From point C, vector **CD** is then drawn in the given direction. When **CD** intersects a vertical drawn downward from **A**, point **D** is established. The magnitude of the force in mast **DA** can now be scaled, as it is shown true length in the front view.
5. The true length of vectors **BC** and **CD** are found by revolution. (These magnitudes could also be found by the auxiliary-view method.)
6. With the correct sense of direction for each vector established, the arrowheads can be transferred to the space diagram to determine the type of load. Vectors **BC** and **CD** are pulling away from the point of concurrency and, thus, are tensile forces. Vector **DA** is pointing toward the concurrent point which indicates a compressive force. Finally, the scaled magnitudes and types of forces are listed in a load table.

11.7.2 Two Unknown Forces Appear Coincident

A structure supporting a 2,000-pound load is shown in Figure 11-19. Bow's notation is used, beginning with the known force **AB** in the front view. Because three unknown forces appear in both the top and front views, the vector polygon for this structure cannot be closed in either view. A new view, 1, in which two of the unknowns appear as an edge, is drawn. Only two unknowns are visible in auxiliary view 1. The vector polygon can then be constructed using view 1 and the horizontal view.

When constructing the vector diagram, the following procedure is used:

1. Remember that the vectors in each view (H and 1) of the vector diagram **must** be constructed parallel to their counterparts in the horizontal and auxiliary views of the space diagram.
2. The known force **AB** is laid out directionally to an arbitrary length in the horizontal and auxiliary views (x_H and x_1). Vector **AB** is rotated until it is parallel to the H/1 fold line (x_{1R}). A true length line Ax_{HR} is seen in the horizontal view. The 2,000-pound magnitude of **AB** is marked off in this true-length view to locate point **B**. Point **B** is projected onto the vector **AB** in both views.
3. Beginning with the auxiliary view of the vector diagram, vector **DA** is drawn through point **A** in the appropriate direction. A line parallel to the edge view of vectors **BC** and **CD** is drawn through point **B**. The intersection of these lines locates point **D**.
4. Through point **A** in the horizontal view, a line parallel to vector **DA** is drawn. Point **D** is established by alignment with the auxiliary view of point **D**.
5. Through point **B** in the horizontal view, a line is drawn parallel to vector **BC**. Through point **D**, a line is drawn parallel to vector **CD**. The intersections of vectors **BC** and **CD** in the horizontal view establishes point **C**, which can now be projected to view 1.
6. Either using rotation, or the auxiliary-view method used in this example, the true magnitudes of vectors **BC**, **CD**, and **DA** must be determined, as well as the compressive or tensile nature of the force. As seen in the load table, members **BC** and **CD** are in compression and member **DA** is in tension.

Figure 11-19. Noncoplanar, concurrent vector system in equilibrium: Two unknown forces as an edge.

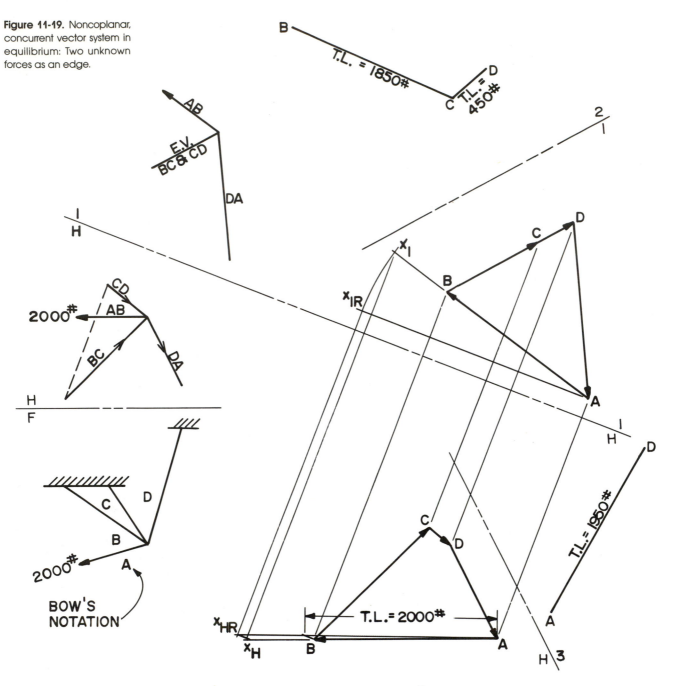

BOW'S NOTATION

SCALE: 1" = 1000 LBS

LOAD TABLE		
AB	2000#	GIVEN
BC	1850#	C
CD	450#	C
DA	1950#	T

CHAPTER PROBLEMS

Problem 1: Determine the resultant force of the concurrent, coplanar force system shown. Use the vector-polygon method to sum these forces. Vector diagram scale: ½" = 100 lb

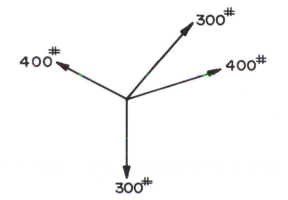

300#

400#

400#

300#

SPACE DIAGRAM

DR. BY:	COURSE & SEC:	SCALE:	DATE:

Problem 2: Find the magnitude and sense of direction of the resultant of the concurrent, coplanar force system shown. Use the vector-polygon method. MAGNITUDE _____ , SENSE

(ANGLE FROM VERTICAL) _____. Vector diagram scale: 1" = 100 lb

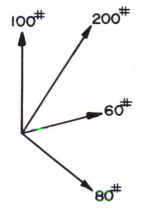

SPACE DIAGRAM

| DR. BY: | COURSE & SEC: | SCALE: | DATE: |

Problem 3: A 350-pound force, inclined at any angle of 30° with the horizontal, is applied to a body resting on the floor. What is the magnitude of the force that tends to move the body along the floor, and of the force which tends to lift the body vertically? Vector diagram scale: 1" = 100 lb

FLOOR

350 LBS

SPACE   DIAGRAM

Problem 4: The pressure angle between a plate cam and its roller follower is 35°, and the pressure is 80 pounds. Resolve the 80-pound force into its vertical lift and its horizontal side thrust components. Vector diagram scale: 1" = 20 lb

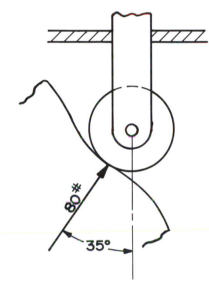

SPACE DIAGRAM

DR. BY:	COURSE & SEC:	SCALE:	DATE:

Problem 5: Three velocity vectors are acting as shown. The true values of the vectors are **AB** = 134 feet per minute, **AC** = 113 feet per minute, and **AD** = 60 feet per minute. Draw a vector diagram and find the value of the resultant velocity vector. Also show the true slope and bearing of the resultant. Vector diagram scale: 1" = 30 ft/min

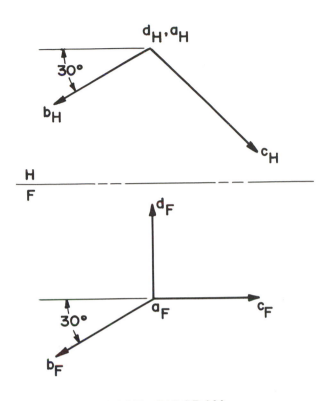

SPACE DIAGRAM

Problem 6: Find the magnitude and sense of direction of the resultant of the given concurrent, noncoplanar vector system. Show the resultant in the space diagram also. The true value of the four forces are **AB** = 500 pounds, **AC** = 385 pounds, **AD** = 470 pounds, and **AE** = 430 pounds. Vector diagram scale: 1" = 300 lb

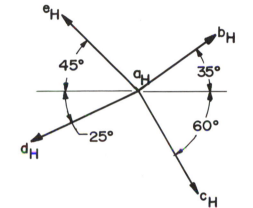

H
F

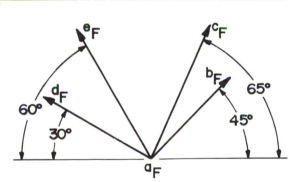

SPACE DIAGRAM

| DR. BY: | COURSE & SEC: | SCALE: | DATE: |

Problem 7: A 750-pound weight is supported by cables. The force acting on one cable is 600 pounds and on the other 500 pounds. Find the angles **A** and **B** indicated for a system in equilibrium. Use Bow's notation to identify the members. Vector diagram scale: 1" = 100 lb

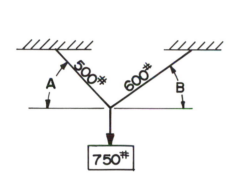

SPACE DIAGRAM

Problem 8: Determine the forces acting in the members **BC** and **CA** for equilibrium. Indicate whether each force is in tension or compression. Vector diagram scale: 1" = 400 lb

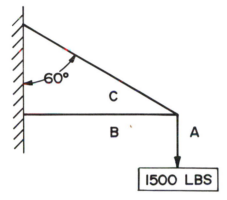

SPACE DIAGRAM

Problem 9: Determine the forces acting in the truss shown below, if the system is in equilibrium. In the space diagram, use Bow's notation to label the members. Draw a Maxwell diagram to solve for the force acting on each member, and designate whether the member is in compression (C) or tension (T). (Use a load table to display your answers.) Maxwell diagram scale: 1" = 500 lb

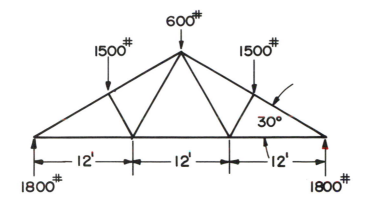

SPACE DIAGRAM

Problem 10: Find the forces acting on the members of the frame shown. Indicate whether forces are compressive (C), or tensile (T). On a C-size (17"×22") drawing sheet, draw the required vector diagram. Show your answers in a load table. Vector diagram scale: 1" = 200 lb

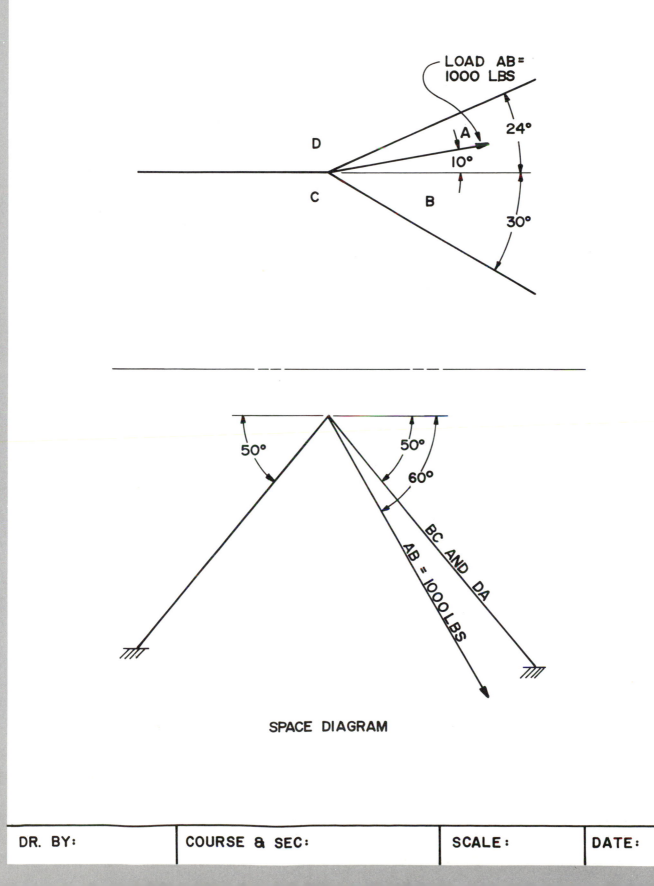

SPACE DIAGRAM

CHAPTER ELEVEN
TEST

1. Define the following terms:

   a. Vector quantity

   b. Coplanar vector system

   c. Nonconcurrent vector system

   d. Resultant

   e. Space diagram

   f. Vector diagram

2. Explain the difference between a resultant and an equilibrant.

3a. List two methods of adding concurrent vectors.

3b. Explain the difference between the two methods listed in 3a.

3c. Are two concurrent vectors necessarily coplanar? Explain.

4. Three concurrent, coplanar vectors are added together and do not form a closed polygon. What does this mean?

5. If a given resultant force is resolved into its horizontal and vertical components, which will be the largest force? Why?

6. When adding noncoplanar vectors, what it the most important consideration when scaling the magnitude of the vectors?

7. Explain Bow's notation.

8. When looking at a vector diagram, how can you tell if the force system is in equilibrium?

9. How do you know if a load is compressive or tensile in nature?

10a. In a nonconcurrent, coplanar vector system, how many unknown vectors can be determined?

10b. How many unknowns are possible in a concurrent, noncoplanar system?

10c. How many unknown vector quantities can you have in any one view of a vector system?

11. What is a Maxwell diagram?

12. How would you determine the magnitudes of the forces in a concurrent, noncoplanar vector system if three unknown forces appeared in the top and front views of the system? (Explain as a general guide to process.)

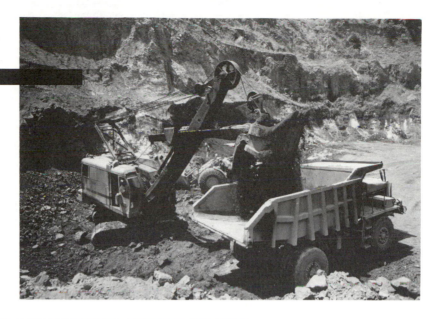

Mining and Civil Engineering Applications

12.1 TERMINOLOGY

The principles of descriptive geometry which have been explained thus far can be particularly useful in mining and civil engineering. When solving mining or highway construction problems, you will work with **topographic maps**, representing the irregularities of the earth's surface in a single view (horizontal). The terrain is shown in the topographic map with **contour lines**. A contour line denotes a series of connected points at a particular elevation. A person walking on a contour line would be on a level path.

The contour line is labeled to indicate its height above sea level. For readability every fifth line is drawn darker and thicker and is called an **index contour line**. In Figure 12-1, the difference in elevation between adjacent contour lines, called the **contour interval**, is 100 feet. The small circle inside the 3,500-foot contour labeled 3,520 feet is the top of the peak. A second peak can be seen to the southeast at 3,350 feet. The two peaks side by side form a **saddle**, which is the low spot between them. Notice that there are small dents in the contour lines. These represent **ridges** and **ravines**. If the contour lines bulge toward lower contour lines, a ridge is present. If the dents bulge toward higher contour elevations, a ravine or water course exists. The spacing of the contour lines indicates the steepness of the ground. A steep slope is indicated by closely spaced contours, whereas greater intervals between contour lines is evidence of a gentle slope.

Below the topsoil and loose rock that covers much of the earth's surface are a series of layers, or **strata**. Originally, the stratified rock, with a few exceptions, formed in continuous horizontal layers. In the years following their formation, geological upheaval caused these layers to become distorted. Today the originally horizontal strata are rarely level and are found in planes over limited areas only.

Within a specific area it is reasonable to assume that a stratum (layer) of rock is uniform in thickness and that it lies between two parallel planes called **upper and lower bedding planes**. Cracks in the rock that often fill with minerals or ores are called **veins**, or **lodes**. Although they vary greatly in shape, veins usually lie between parallel planes. In the illustration seen in Figure 12-2, inclined parallel planes of sand, coal, shale, and limestone are shown.

375

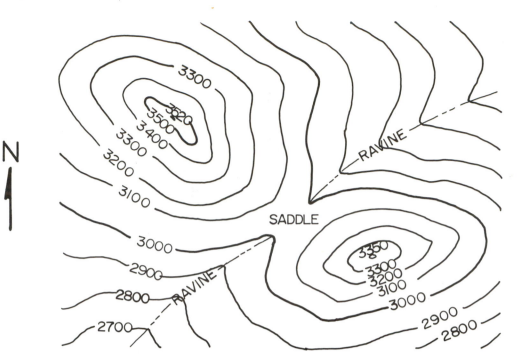

Figure 12-1. A topographic map.

Often these parallel strata are interrupted by fractures where one side of the bed shifts in relation to the other. This is called a **fault**; and the plane of the fracture is called the **fault plane**. When the fault plane intersects the earth's surface it is called a **fault-plane outcrop**.

The term **outcrop** is also used to describe where a stratum of rock, or a vein of ore intersects the earth's surface. In Figure 12-2 a coal stratum intersects the hillside on the left. By noting these surface conditions, from drilled test holes, and from other geological tests, a mining engineer can estimate the approximate position of an underlying stratum. With data compiled, the engineer can determine the location and depth of three or more points on a bedding plane. Since three points form a plane, the location of the whole stratum can be established, and the future working of the area can be planned.

12.2 LOCATION AND ANALYSIS OF A STRATUM

Figure 12-3(a) is of a portion of a topographic map. Points **R**, **S**, and **T** in the upper bedding plane are given, as well as point **P** on the lower bedding plane. The elevation of each point is given in parentheses. Point **S**, at an elevation of 1,300 feet, lies on the 1,300-foot contour line, and thus is a point on the surface (an outcrop). Boreholes have been drilled to locate points **R** (950 feet) and **T** (750 feet) on the upper bedding plane, and also point **P** (725 feet) on the lower bedding plane.

In mining it is common practice to describe the location of a stratum using strike and dip. The **strike** is the bearing of a horizontal line in the plane of the stratum, customarily measured from north, as N65°E which is seen in Figure 12-3(b). The **dip** is the slope angle of the plane of the stratum, given in conjunction with the general direction of the downward slope of the plane. This dip direction is always perpendicular to the strike line, as shown in Figure 12-3(b).

Examine Figure 12-3(b), and follow the procedure for determining the strike, dip, and thickness listed below:

1. The horizontal view is a copy of the map in Figure 12-3(a) with the contour lines deleted. The front has been drawn to show the elevations of points **P**, **R**, **S**, and **T**.
2. A horizontal line, **RX**, is constructed in the front view ($r_F x_F$), and projected into the top view at $r_H x_H$. The strike of plane **RST** is the bearing of this line, and is lettered along the line as shown (N65°E).

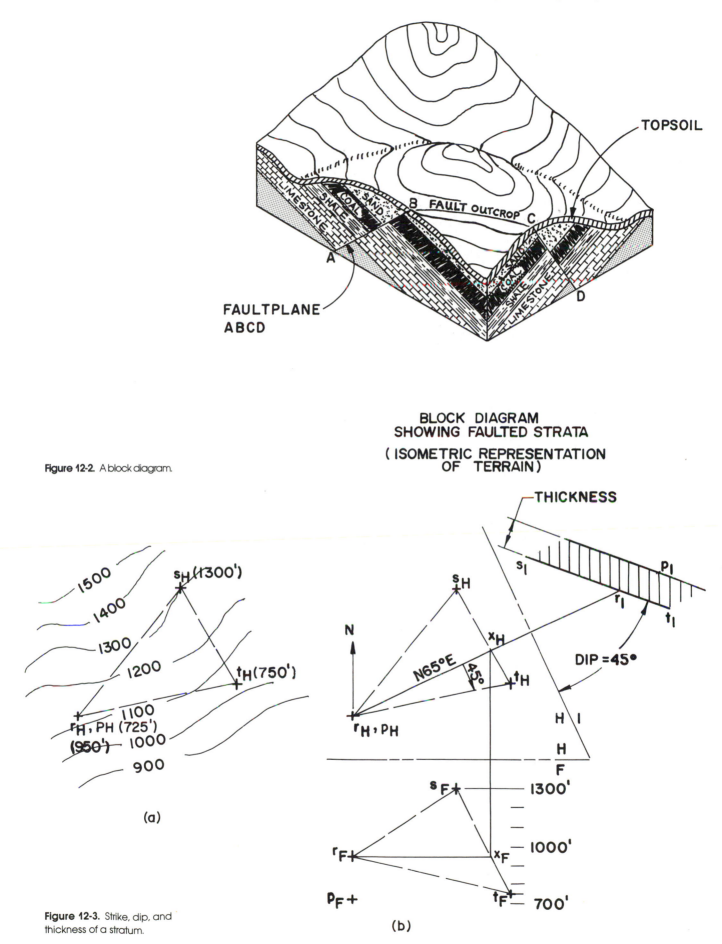

Figure 12-2. A block diagram.

BLOCK DIAGRAM
SHOWING FAULTED STRATA
(ISOMETRIC REPRESENTATION
OF TERRAIN)

Figure 12-3. Strike, dip, and thickness of a stratum.

(a)

(b)

3. To find the dip, an edge view of the stratum must be found. Fold line H/1 is drawn perpendicular to $r_H x_H$, and an edge view of the stratum is seen in view 1. Here the slope angle of 45° is the dip. The dip shown in the horizontal view as a short arrow drawn perpendicular to the strike line, and pointing toward the lower side of the plane. The dip would be verbally described as 45° downward in a generally south-easterly direction.

4. It is customary to assume that the upper and lower bedding planes are parallel. Thus the edge view of the lower bedding planes is drawn through p_1, and parallel to the edge view of **RST** (r_1, s_1, t_1). The thickness of the stratum is measured on a perpendicular between the edge views.

12.3 OUTCROP OF A STRATUM OR VEIN

When a stratum or vein intersects the earth's surface, it is called an **outcrop**. When the position of the stratum has been determined, then the probable outcrop can be found, and the most economical mining of the vein can be planned. The auxiliary view showing the vein as an edge will show the contour lines as straight horizontal lines. This view shows clearly where the contour lines intersect the edges of the vein. These points of intersection can be projected to the top (map) view and connected to show the outcrop.

In Figure 12-4 three points, **X**, **Y**, and **Z** on the upper bedding plane, and point **W**

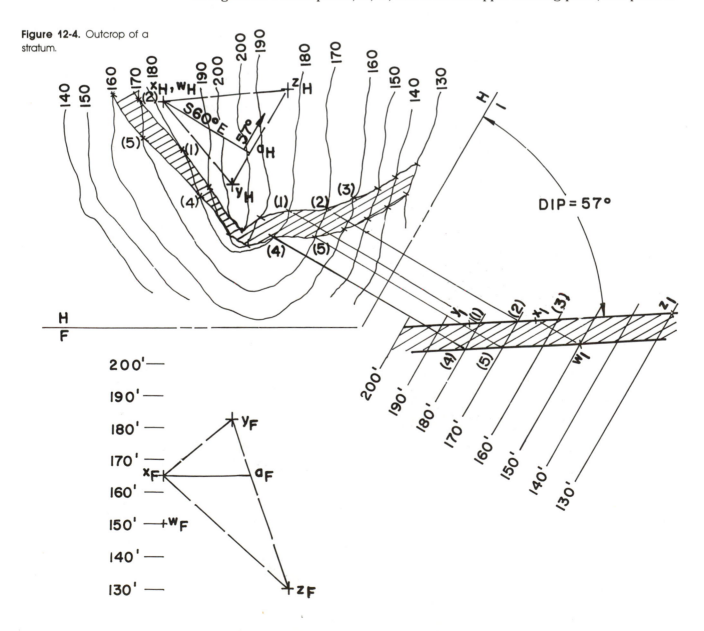

Figure 12-4. Outcrop of a stratum.

on the lower bedding plane, are given in both the horizontal and front views. The following procedure was followed to determine the outcrop (the shaded area):

1. Since the locations of points **X**, **Y**, and **Z** on the upper bedding plane are known, the strike, dip, and thickness can be determined by the method discussed in section 12.2. Line **AX** shows the strike to be S60°E. In view 1, the edge view is shown indicating a 57° dip in a northeasterly direction. Point **W** is directly below **X** and on the lower bedding plane. Thus a line through w_1 parallel to plane **XYZ** establishes the lower bedding plane. The thickness of the bedding plane is measured on a perpendicular between the edge views.

2. In view 1, the point where the stratum intersects the 180-foot contour line is marked **(1)**. This point of intersection is projected to the horizontal view, where it twice crosses the 180-foot contour line as shown. Follow the same procedure to locate point of intersection **(2)**. The point twice intersects the 170-foot contour line in the horizontal view. Do the same for each consecutive point of intersection on the upper bedding plane.

3. The lower outcrop line is determined in the same manner except that contour intersections points are taken on the lower bedding plane. Note that point **(4)** on the 180-foot contour line in the lower plane is projected to the horizontal view, intersecting the 180-foot contour as shown. By continuing this procedure, points of outcrop can be located on each successive contour line to establish the lower outcrop line.

4. When both outcrop lines have been determined, the area is shaded in the horizontal view for readability.

12.4 CUTS AND FILLS

Highways constructed through irregular terrain involve the principles of the intersection of a line (contour line) and a plane (the earth's surface), known as **cut and fill**.

12.4.1 Cuts and Fills Along a Level Road

A proposed highway, shown as centerline **AB** in Figure 12-5, is to be 80 feet wide at a constant elevation of 120 feet. Earth cuts are to be made at a slope of 1:1 (one unit horizontally to one unit vertically). Earth fills are to be made at a slope of 1-½:1 (one and one-half units horizontally to one unit vertically).

The problem is to determine the location of the required cuts and fills for the example shown in Figure 12-5. The procedure for doing so is as follows:

1. The horizontal view shows the roadway on a given contour map. The front view shows the profile, a vertical section of the terrain along the centerline of the road.

2. An edge view (point view of centerline **AB**) of the highway is shown in view 1. Here the cut and fill ratios are drawn. The cut ratio of 1:1 is drawn upward, toward fold line H/1. The fill ratio of 1-½:1 is drawn downward, away from fold line H/1. In this view, the contour lines appear as straight horizontal lines.

3. Begin with the cut and project the points of intersection between the cut slope and the 170-foot elevation onto the 170-foot contour line in the horizontal view. The remaining points of intersection between the cut and the contour lines in view 1 are projected to their respective contour lines in the horizontal view to determine the limits of the cut. The line indicating the limits of the cut is labeled **top of the cut**.

4. Again, in view 1, you will see the points where the fill slope intersects the various contour elevation lines. These points are projected onto their respective contour lines in the horizontal view denoting the limits of the fill. The line indicating the limit of the fill is labeled **toe of the fill**.

12.4.2 Cuts and Fills Along a Grade Road

If the highway is sloping, a slightly different procedure must be followed. Line **AB**, in Figure 12-6, is the centerline of a proposed 50-foot wide highway having a 10 percent grade.

At station 1 + 00 the elevation of the highway is 90 feet. Cuts are to be made at a

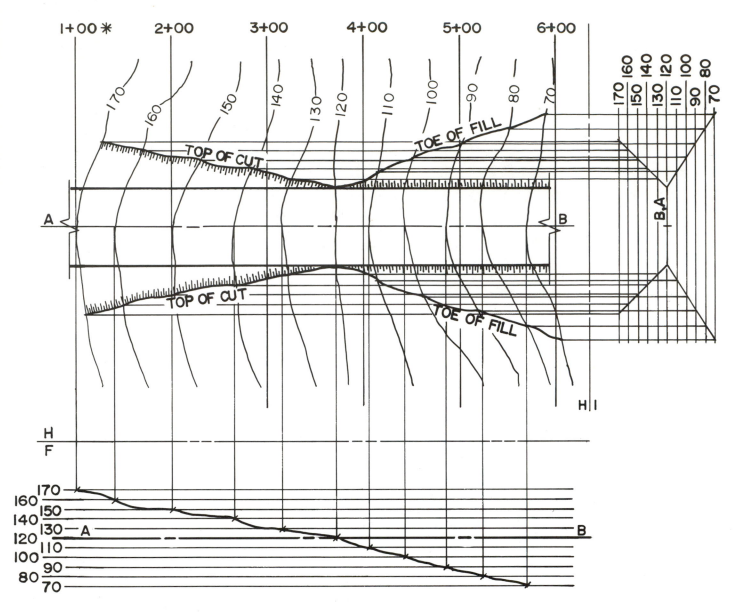

Figure 12-5. Cuts and fills along a level road.

✳NOTATION USED IN SURVEYING. STATION 2+00 IS 100 FEET FROM STATION 1+00. A POINT HALF-WAY BETWEEN THESE TWO STATIONS WOULD BE MARKED 1+50.

1:1 slope, while fills are to be made at a 1-½:1 slope. At station 2 + 00 the elevation of the highway is 100 feet. Since the highway has a grade of 10 percent in 100 feet of horizontal run, the elevation of the highway will rise 10 feet from 90 to 100 feet.

To find the required limits of cut and fill, the following procedure has been followed:

1. Begin at station 1 + 00 on the north edge of the road. Lay off a distance of 10 feet from the edge of the road. The elevation at this point, **X**, will be 100 feet. This is true because the contour lines in the vicinity of stations 1 + 00 and 2 + 00 are generally higher than the elevation of the road, thus cuts will be required. Since the cut slope is 1:1, a horizontal distance of 10 feet will mean a 10-foot increase in elevation. Lay off additional lines at intervals of 10 feet and mark their elevations.

2. At station 2 + 00 lay off a distance of 10 feet from the edge of the road. At this point the elevation will be 110 feet, because the road at this station is 100 feet in

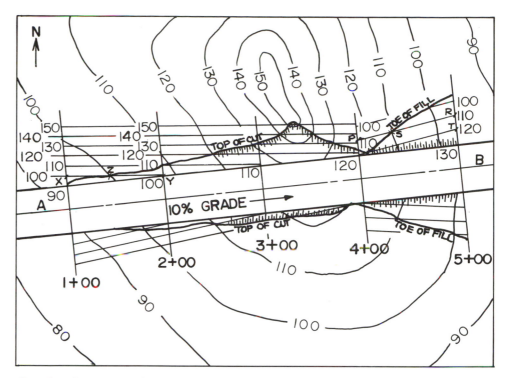

Figure 12-6. Cuts and fills along a grade road.

elevation. Lay out additional lines at 10-foot intervals as was done at station 1 + 00.

3. Notice that point **X** at station 1 + 00 is the same elevation as is the road at station 2 + 00 (point **Y**). The line joining these points is a contour line on the sloping 1:1 surface. This contour line intersects the earth's contour (elevation 100 feet) at point **Z**. This is a point common to the earth's surface and to the cut plane (1:1) surface.

4. By connecting similar lines on the sloping surface, and by determining their intersections with the respective earth-contour lines, the top of the cut can be found. Notice that a similar process was followed on the south side of the road.

5. In the general vicinity of station 4 + 00, observe that the contour lines are now lower than the proposed highway. This area will require fill. If we lay off a distance of 15 feet from the north edge of the road, the elevation at point **P** will be 110 feet. If we lay off another 15 feet from point **P**, the elevation will be 100 feet, and so forth. Remember that the slope of the fill is 1-½:1, thus a 15-foot horizontal distance means a 10-foot drop in elevation. The road at station 4 + 00 is 120 feet.

6. Move to station 5 + 00 where the road elevation is 130 feet. Again lay off a distance of 15 feet from the north edge of the road, establishing a point at an elevation of 120 feet (**T**). If we lay off another 15 feet from **T** to **R**, the elevation of **R** will be 110 feet. Repeat this process to determine additional contour lines on the sloping surface, 1-½:1. Connect a line from the road elevation of 120 feet at station 4 + 00 to point **T** at station 5 + 00. Connect another line from point **P** to point **R** (both at 110-foot elevations). Notice that this line intersects the 110-foot contour line at point **S**. Additional points are located in the same manner. The line joining these common points forms the toe of the fill. This same procedure is followed for the fill on the south side of the highway.

This discussion of the application of descriptive geometry to civil engineering problems is complete for the purposes of this text. Yet it might be helpful for you to know at least a few of the steps that might follow the determination of cuts and fills. It would then be possible to draw cross sections of a highway. The areas of the cross sections can be determined, and with the distances between sections known, the volumes of cuts and fills can be calculated. A large amount of these kinds of calculations are currently being made using computers as are the designs and drawings for many civil engineering projects.

CHAPTER PROBLEMS

Problem 1: Points **R, S,** and **T** are on the upper bedding plane of a stratum of coal. Drilling has also determined a point **P** on the lower bedding plane. Find the strike, dip, and thickness of this stratum. Scale: 1" = 100'

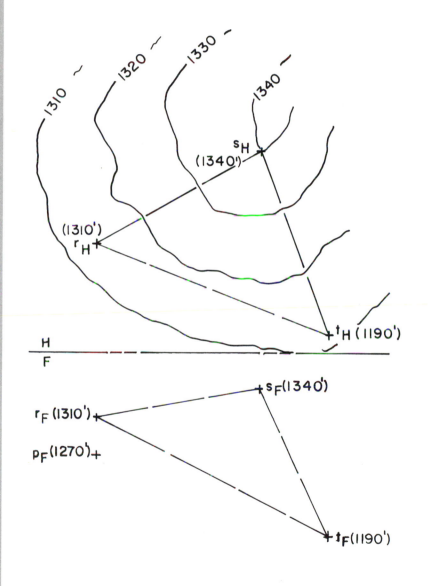

1310 1320 1330 1340

^sH
(1340')

(1310')
^rH

^tH (1190')

H
―――
F

s_F(1340')

r_F (1310')

p_F(1270')

t_F(1190')

| DR. BY: | COURSE & SEC: | SCALE: | DATE: |

Problem 2: While exploring for oil-bearing sand, a company drills at point **X** and hits oil-bearing sand at 750 feet below the surface. Another drilling at point **Y** strikes the same sand at 1660 feet below the surface. A third drilling at point **Z** hits the stratum of sand at 1790 feet below the surface. All three drillings indicate a stratum thickness of 40 feet. Find the strike and dip of this oil-bearing sand layer. Show the upper and lower beddings planes in your solution. Scale: 1" = 500'

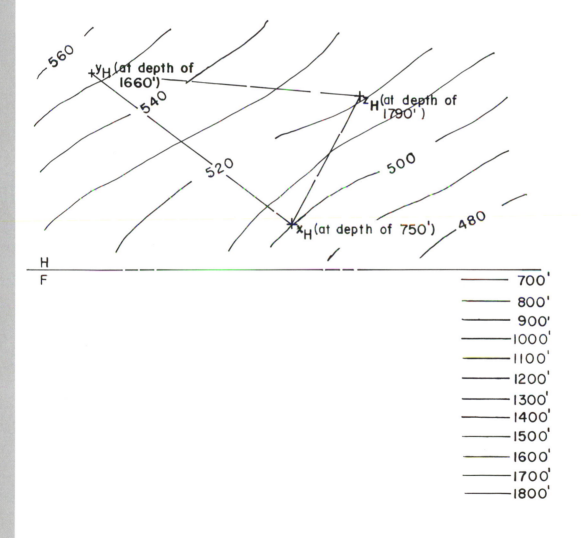

- 560
- +Y_H (at depth of 1660')
- 540
- +Z_H (at depth of 1790')
- 520
- 500
- +X_H (at depth of 750')
- 480

H
F

———	700'
———	800'
———	900'
———	1000'
———	1100'
———	1200'
———	1300'
———	1400'
———	1500'
———	1600'
———	1700'
———	1800'

Problem 3: From the bench mark (B.M.), the locations of three vertical boreholes are shown. Distances along the given bearings are map distances from the bench mark. The drillings have established points **X**, **Y**, and **Z** on the upper bedding plane of a vein of ore. All three drillings indicate a lower bedding plane 50 feet below the upper. Find the strike, dip, and thickness of the vein. Your finished drawing should be done on a B-size drawing sheet. Scale: 1" = 200'

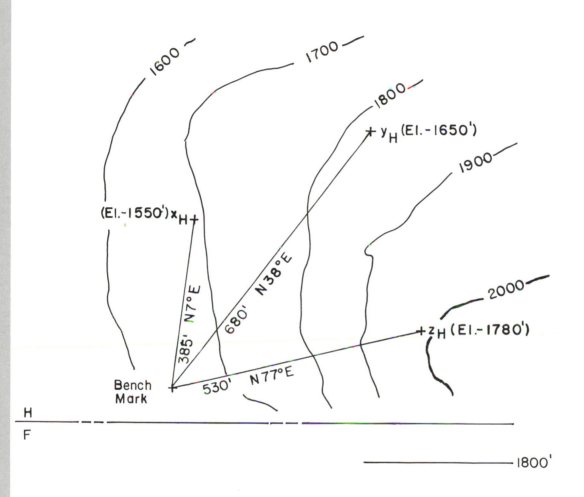

| DR. BY: | COURSE & SEC: | SCALE: | DATE: |

Problem 4: Points **A** and **C** are on the outcrop of the upper bedding plane of a vein of low-grade ore. Continued drilling at **A** has intersected the lower surface of the vein 20 feet below **A**. The strike of the vein is N75°W from point **A**. Find the dip, thickness, and probable outcrop of the upper and lower planes. Indicate the outcrop with crosshatched lines. Trace this map and show your solution on a B-size drawing sheet. Scale: 1" = 40'

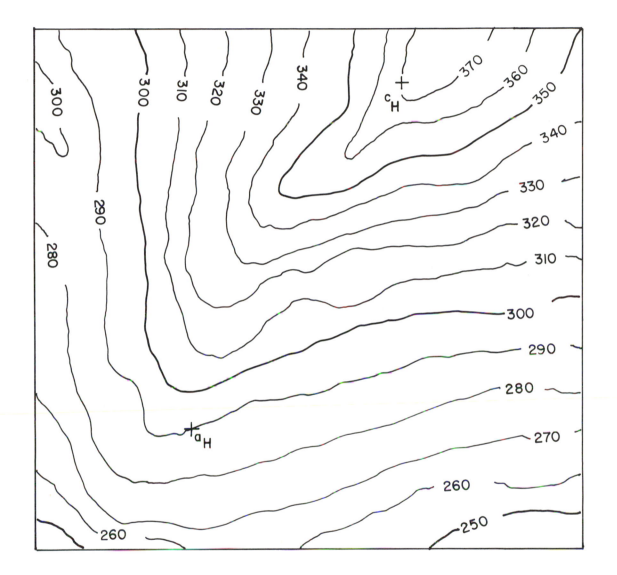

DR. BY:	COURSE & SEC:	SCALE:	DATE:

Problem 5: Line **XY** is the centerline of an access road (30 feet wide) on the top of an earth dam. The elevation of the road is 430 feet. Trace the given map on a B-size drawing sheet. Show the cut and fill on both sides of the dam. The cut has a 2:1 slope, and the fill a 3:1 slope. Indicate the cuts and fills in accordance with the practice shown in Figure 12-6 in this text. Draw a front view showing the profile of the terrain along the center of the road. Scale: 1″ = 100′

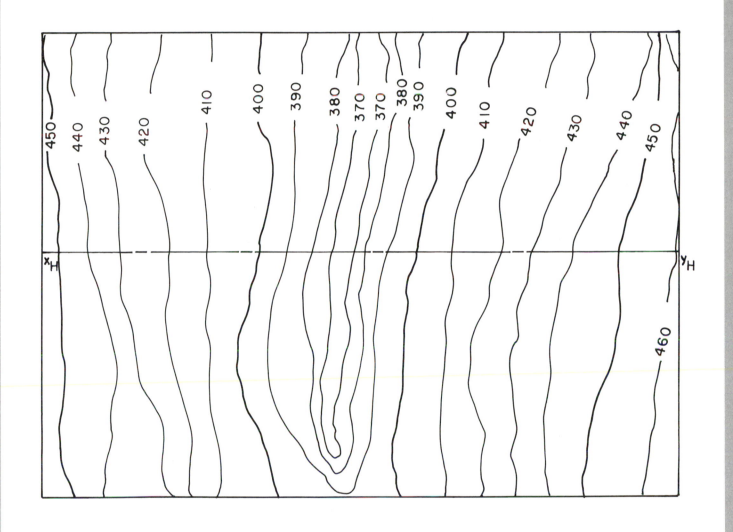

| DR. BY: | COURSE & SEC: | SCALE: | DATE: |

Problem 6: Line **AB** is the centerline of a 60-foot wide, level roadway that has an elevation of 210 feet. Trace this map on a B-size drawing sheet. Plot the lines of the cut and fill on the map. The cut is to be made at a 1:1 slope, and the fill at a 2:1 slope. Identify the top of the cut and the toe of the fill, indicating cuts and fills as shown in Figure 12-6 in this text. Scale: 1" = 100'

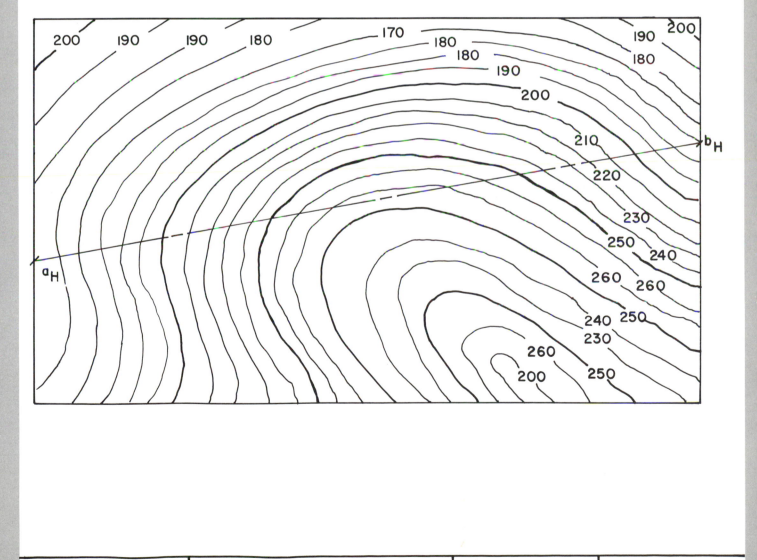

Problem 7: Line **AB** is the centerline of a 50-foot wide roadway. The road has a 5 percent grade up from station 1 + 00. In the front view draw the profile of the terrain along the centerline of the road. Plot the top of the cut and the toe of the fill. The cut slope is 1:1, and the fill slope is 2:1. Indicate the cuts and fills as shown in Figure 12-6 in this text. Scale: 1" = 100'

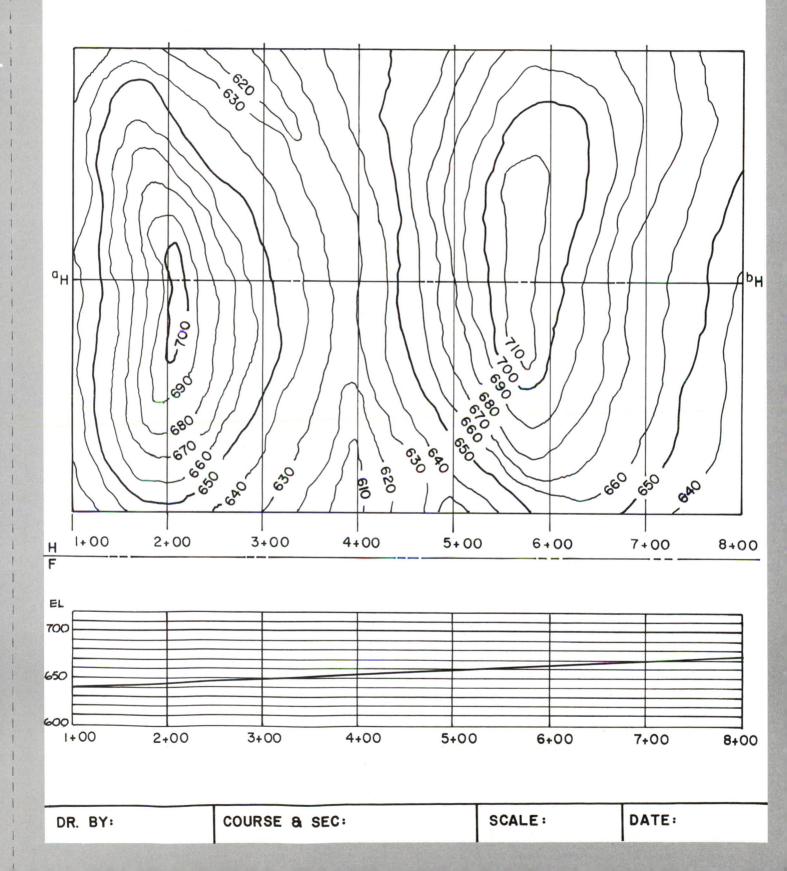

| DR. BY: | COURSE & SEC: | SCALE: | DATE: |

Problem 8: Locate the top of the cut and the toe of the fill for a 50-foot wide road, the centerline of which runs N75°W from point **Y** (station 1+00). The road has a 5 percent grade up from station 1+00, which is at the 150-foot elevation. The cut slope is 1:1, and the fill slope is 2:1. Indicate the cuts and fills as in previous problems. Scale: 1" = 100'

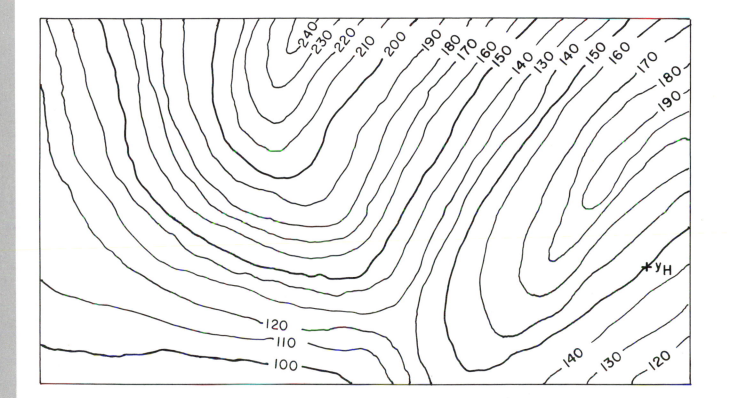

CHAPTER TWELVE
TEST

1. What is a topographic map?

2. What are contour lines?

3. How can you tell the difference between ridges and ravines?

4. How is the top of a peak or hill noted on a topographic map?

5. Define the following:
   a. Stratum of rock

   b. Bedding planes

   c. Fault

   d. Outcrop

   e. Strike

   f. Dip

6. Explain how you would determine the thickness of a vein of ore.

7a. In which view would you see the strike of a plane? Why?

7b. In which view would you see the dip of a plane? Why?

8. The outcrop of a vein of ore represents the intersection of _____ and _____ .

9. Explain the term *cuts and fills*.

10. Explain a fill ratio of 2:1 (you may use a sketch to help illustrate your explanation).

11. What do the terms *top of the cut* and *toe of the fill* mean?

12. Each point on a top-of-the-cut line or the toe-of-the-fill line is a point common to which two surfaces?

GLOSSARY

Auxiliary View: An auxiliary view is a view on any projection plane other than a primary, or principal, projection plane.

Auxiliary-elevation view: An auxiliary-elevation view is an elevation view which is not one of the principal elevation views. Auxiliary-elevation projection planes are always perpendicular to the horizontal projection plane.

Bearing of a line: The bearing of a line is the angular relationship of the horizontal projection of the line relative to the compass, expressed in degrees.

Bedding planes: A stratum of rock in the earth's surface lies between two parallel planes, called bedding planes.

Bow's notation: Bow's notation is a system of notation used to label a vector system. A letter is given to the space on each side of a vector, and each vector is then identified by the two letters on either side of it, reading in a clockwise direction.

Concurrent vector system: In a concurrent vector system, all vectors act on a common point.

Contour line: A contour line denotes a series of connected points at a particular elevation.

Coplanar vector system: In a coplanar-vector system, all vectors lie in the same plane.

Descriptive geometry: Descriptive geometry is the theory of orthographic projection. It can also be defined as a graphical method of solving solid (or space) analytic geometry problems.

Dihedral angle: The dihedral angle is the angle that is formed by two intersecting planes.

Edge view of a plane: An edge view of a plane is one in which the given plane appears as a straight line. An edge view of a plane may be found by viewing as a point a true length line in the plane.

Elevation view: An elevation view is a view in which the lines of sight are level. Principal elevation views are front, left profile, right profile, and rear.

Equilibrant: The equilibrant is a vector that is equal in magnitude to the resultant, and has the opposite direction and sense.

Equilibrium: When a vector system has a resultant of zero, then the system is said to be in equilibrium.

Fold (reference) line: A fold line is a line of intersection between two projection planes.

Foreshortened line: A line which appears shorter than its actual length.

Frontal line: A frontal line is a line which is parallel to the front projection plane, and its projection will be true length in the front view.

Grade of a line: The grade of a line is another way to describe the inclination of a line in relation to the horizontal plane. The precent grade is the vertical rise divided by the horizontal run multiplied by 100.

Horizontal line: A horizontal line is a line which is parallel to the horizontal plane, and its projection will appear true length in the top view.

Horizontal, or top, view (horizontal projection): A horizontal view is an orthographic view for which the lines of sight are vertical and for which the projection plane is level.

Inclined auxiliary view: An inclined auxiliary view is one in which the lines of sight of the observer are neither vertical nor horizontal.

Intersecting lines: When lines are intersecting, the point of intersection is a point that lies on both lines.

Line of sight: A line of sight is an imaginary straight line from the eye of the observer to a point on the object being observed. All lines of sight for a particular view are assumed to be parallel and are perpendicular to the projection plane involved.

Maxwell diagram: A Maxwell diagram is a combination vector diagram used to analyze the forces acting in a truss.

Noncoplanar vector system: In a noncoplanar vector system, the given vectors lie in different planes.

Noncurrent vector system: In a noncurrent vector system, all vectors do not act on a common point.

Normal plane: A normal plane is a plane surface which is parallel to any one of the primary projection planes.

Oblique line: An oblique line is a straight line which is not parallel to any of the six principal planes.

Oblique plane: An oblique plane is inclined to all of the principal projection planes.

Orthographic projection (drawing): Orthographic projection means right-angle projection. It is a method of drawing which uses parallel lines of sight at right angles (90°) to a projection plane.

Outcrop: The term outcrop is used to describe where a stratum of rock or view of ore intersects the earth's surface.

Parallel lines: Lines which are parallel are an equal distance from each other throughout their length.

Perpendicular lines: Lines are perpendicular when there is a 90° angle between them.

Piercing point: A piercing point is a point where a particular line intersects a plane.

Pitch: Pitch is another measure of slope and is expressed as a ratio of vertical rise to 12 feet of horizontal span (run).

Plane: A plane is a surface which is not curved or warped. It is a surface in which any two points may be connected by a straight line, and the straight line will always lie completely within the surface.

Point view of a line: A point view of a line is the view seen through a projection plane drawn perpendicular to the true length of the given line.

Primary auxiliary view: A primary auxiliary view is any auxiliary view that is adjacent to a primary, or principal, plane.

Primary, or principal, view: The view represented by the six sides of the orthographic box. They include the top, front, right profile, left profile, rear, and bottom views.

Profile line: A profile line is one that is parallel to the profile projection plane and its projection appears true length in the profile view.

Projection line: A projection line is a straight line at 90° to the fold line, which connects the projection of a point in a view to the projection of the same point in the adjacent view.

Projection plane: A projection plane is an imaginary surface on which the view of the object is projected and drawn. This surface is imagined to exist between the object and the observer.

Ravine: If contour lines bulge toward higher contour elevations, a ravine, or water course, exists.

Resultant: The resultant is the sum of the vector system. The resultant has the same effect as the total of the individual vectors.

Revolution: Revolution is an alternate method for solving descriptive geometry problems in which the observer remains stationary and the object is moved to obtain the various views of it.

Ridge: If contour lines on a topographic map bulge toward lower contour elevations, a ridge is present.

Secondary auxiliary view: A secondary auxiliary view is an auxiliary view that is adjacent to a primary auxiliary view.

Slope of a line: The slope of a line is the angle in degrees that the line makes with a horizontal plane.

Slope of a plane: The slope of a plane, also called the dip angle, is the angle in degrees that the edge view of a plane makes with a horizontal plane.

Space diagram: A space diagram is a drawing of a vector system, showing the correct direction and sense, but not drawn to scale.

Stratum of rock: Below the topsoil and loose rock which cover the earth's surface are a series of parallel layers of stratified rock. Each layer is called a stratum.

Surface development: When the surfaces of a solid are unfolded, or unrolled onto a flat plane, a surface development results. The important characteristics of surface developments is that all lines appear true length in the development.

Topographic map: A topographic map is a map that represents the irregularities (contours) of the earth's surface in a single horizontal view.

True length line: A line that appears in its actual and true length.

True shape of a plane: The true shape of a plane is the actual shape and size of a plane surface. It can be seen on a projection plane that is parallel to the plane surface.

Vector diagram: A vector diagram is a drawing of the vector system in which the vectors are drawn with the correct magnitude and sense, and to scale.

Vector quantity: A vector quantity is one that requires both magnitude and direction for its complete description.

Vector system: A vector system is created when two or more vectors act on a object.

Vein: Cracks in the earth's strata often fill with minerals, or ores. These cracks are called veins or lodes.

Delmar Publishers Inc. and the author would like to thank the following companies and people for providing chapter opening photographs:

Chapter 1 Duramic Products Inc.
Chapter 2 Brian Feeney
Chapter 3 Bethlehem Steel Corporation
Chapter 4 Cincinnati Milacron
Chapter 5 Portland Cement Association
Chapter 6 Bethlehem Steel Corporation
Chapter 7 Madsen Construction
Chapter 8 Portland Cement Association
Chapter 9 Madsen Construction
Chapter 10 Majestic Company, Division of American Standard
Chapter 11 Bethlehem Steel Corporation
Chapter 12 Aluminum Company of America